Agisilaos Papadogiannis

Advanced Coordinated Multi-Point Techniques for Wireless Networks

Agisilaos Papadogiannis

Advanced Coordinated Multi-Point Techniques for Wireless Networks

Networks of the future

Presses Académiques Francophones

Mentions légales / Imprint (applicable pour l'Allemagne seulement / only for Germany)
Information bibliographique publiée par la Deutsche Nationalbibliothek: La Deutsche Nationalbibliothek inscrit cette publication à la Deutsche Nationalbibliografie; des données bibliographiques détaillées sont disponibles sur internet à l'adresse http://dnb.d-nb.de.
Toutes marques et noms de produits mentionnés dans ce livre demeurent sous la protection des marques, des marques déposées et des brevets, et sont des marques ou des marques déposées de leurs détenteurs respectifs. L'utilisation des marques, noms de produits, noms communs, noms commerciaux, descriptions de produits, etc, même sans qu'ils soient mentionnés de façon particulière dans ce livre ne signifie en aucune façon que ces noms peuvent être utilisés sans restriction à l'égard de la législation pour la protection des marques et des marques déposées et pourraient donc être utilisés par quiconque.

Photo de la couverture: www.ingimage.com

Editeur: Presses Académiques Francophones est une marque déposée de
Südwestdeutscher Verlag für Hochschulschriften GmbH & Co. KG
Heinrich-Böcking-Str. 6-8, 66121 Sarrebruck, Allemagne
Téléphone +49 681 37 20 271-1, Fax +49 681 37 20 271-0
Email: info@presses-academiques.com

Produit en Allemagne:
Schaltungsdienst Lange o.H.G., Berlin
Books on Demand GmbH, Norderstedt
Reha GmbH, Saarbrücken
Amazon Distribution GmbH, Leipzig
ISBN: 978-3-8381-8940-6

Imprint (only for USA, GB)
Bibliographic information published by the Deutsche Nationalbibliothek: The Deutsche Nationalbibliothek lists this publication in the Deutsche Nationalbibliografie; detailed bibliographic data are available in the Internet at http://dnb.d-nb.de.
Any brand names and product names mentioned in this book are subject to trademark, brand or patent protection and are trademarks or registered trademarks of their respective holders. The use of brand names, product names, common names, trade names, product descriptions etc. even without a particular marking in this works is in no way to be construed to mean that such names may be regarded as unrestricted in respect of trademark and brand protection legislation and could thus be used by anyone.

Cover image: www.ingimage.com

Publisher: Presses Académiques Francophones is an imprint of the publishing house
Südwestdeutscher Verlag für Hochschulschriften GmbH & Co. KG
Heinrich-Böcking-Str. 6-8, 66121 Saarbrücken, Germany
Phone +49 681 37 20 271-1, Fax +49 681 37 20 271-0
Email: info@presses-academiques.com

Printed in the U.S.A.
Printed in the U.K. by (see last page)
ISBN: 978-3-8381-8940-6

TELECOM
ParisTech

EURECOM
Sophia Antipolis

THESIS
In Partial Fulfillment of the Requirements
for the Degree of Doctor of Philosophy
from TELECOM ParisTech

Specialization: Electronics and Communications

Agisilaos Papadogiannis

Systems and Techniques for Multicell-MIMO and Cooperative Relaying in Wireless Networks

Thesis defended on the 11th of December 2009 before a committee
composed of:

President	Prof. Jean-Claude Belfiore, TELECOM ParisTech, France
Reviewers	Prof. Mérouane Debbah, SUPELEC, France
	Prof. Gema Piñero, Universidad Politechnica de Valencia, Spain
Examiners	Dr. Rodrigo de Lamare, University of York, UK
	Dr. Eric Hardouin, Orange Labs, France
Thesis supervisor	Prof. David Gesbert, EURECOM, France

TELECOM
ParisTech

EURECOM
Sophia Antipolis

THESE
présentée pour obtenir le grade de

Docteur de TELECOM ParisTech

Spécialité: Electronique et Communications

Agisilaos Papadogiannis

Systèmes et Techniques MIMO Coopératif dans les Réseaux Multi-cellulaires Sans Fil

Thèse soutenue le 11 décembre 2009 devant le jury composé de:

Président	Prof. Jean-Claude Belfiore, TELECOM ParisTech, France
Rapporteurs	Prof. Mérouane Debbah, SUPELEC, France
	Prof. Gema Piñero, Universidad Politechnica de Valencia, Espagne
Examinateurs	Dr. Rodrigo de Lamare, University of York, Royaume-Uni
	Dr. Eric Hardouin, Orange Labs, France
Directeur de thèse	Prof. David Gesbert, EURECOM, France

To my parents, my brother and Catherine.

ii

Acknowledgements

This thesis has been pursued with the Research division of France Telecom (Orange Labs), Paris, through a *CIFRE*[1] agreement, thus I would like to acknowledge the contribution and support of France Telecom.

I sincerely thank my academic supervisor Prof. David Gesbert for his invaluable guidance, insight and support; his aid has been essential in all the stages of this thesis. I am very thankful to my industrial supervisor Dr. Eric Hardouin for his continuous support, his cordial collaboration and for providing me with the opportunities to develop my skills. I would also like to thank from France Telecom, Dr. Ahmed Saadani for his very constructive and fruitful collaboration in the subject of cooperative communications and my team manager Eric Njedjou Ntonfo for the endorsement of my work and for helping me with all the practical issues I encountered. I am especially grateful to my project manager Alexandre Gouraud for his constant support related to all issues of my professional and personal life.

I would like to specially thank my colleague and friend Hans Jørgen Bang for inviting me to work with him in Oslo and for co-authoring some interesting papers. I am also very thankful to my co-authors and friends Giorgos Alexandropoulos and Muhammad Farukh Munir. In addition I would like to thank all my colleagues at EURECOM for creating a nice and friendly atmosphere. Special thanks go to Daniel Câmara, to Mustapha Amara and to the friends from my office.

Furthermore, I warmly thank the members of my Ph.D defense committee, Professors Jean-Claude Belfiore, Mérouane Debbah, Gema Piñero and Rodrigo de Lamare for their valuable feedback and their very constructive comments on my thesis and its future perspectives.

Finally I would like to express my deepest gratitude to my family for believing in me, inspiring me with great values and motivating me to pursue

---

[1]Convention Industrielle de Formation par la Recherche/Industrial Agreement for Training through Research

ness ...

my dreams. I am especially indebted to my father and to Catherine for their unyielding support and ample understanding during the course of this thesis.

Abstract

The constantly increasing demand for wireless services, the scarcity of radio spectrum and the characteristics of the global wireless market, necessitate that future wireless systems (Fourth Generation Mobile - 4G) provide higher peak data rates and better QoS, especially for the cell-edge users. Furthermore it is essential that they achieve high spectral efficiencies and they are easily deployed. In order to be able to accomplish these objectives, wireless systems need to incorporate technologies that increase the cell throughput without increasing spectral consumption.

A very promising technique that can achieve the aforementioned targets is Multicell Cooperative Processing (MCP) or Multicell-MIMO. MCP has the potential to mitigate Inter-Cell Interference (ICI) and augment data rates without sacrificing additional spectrum but at the cost of some overhead and complexity. According to the concept of clustered MCP proposed in this thesis, Base Stations (BSs) are grouped into cooperation clusters, each of which contains a subset of the network BSs. The BSs of each cluster exchange information and jointly process signals as they form virtual antenna arrays distributed in space. In these systems, each user receives useful signals from several BSs and therefore the notion of a cell transcends the one of the conventional cellular systems. Although Multicell-MIMO is a technique that can help meet a lot of the challenges towards 4G systems, it has some intrinsic drawbacks that need to be addressed in order for it to be brought into practice; this is the main focus of the present thesis.

Firstly the problem of how to optimally form BS cooperation clusters of limited size has been investigated. MCP's overheads are proportional to the size of cooperation clusters, therefore this size should be kept limited. The straightforward solution of forcing neighboring BSs to collaborate provides limited gains. In this thesis it is proposed that the BSs which interfere the most with each other should cooperate rather the ones that are in close proximity. This is shown to lead to significant spectral efficiency gains while cluster sizes are kept very small.

The typical centralized architectural conception for MCP entails that the
BSs of each cooperation cluster should be inter-connected through a control unit and exchange Channel State Information (CSI). This conception
impedes the deployment of MCP systems as it implies additional infrastructural costs. In this thesis a new decentralized framework has been proposed
that allows the incorporation of MCP by the conventional cellular systems
with very few changes upon their architecture. Mobile Stations (MSs) feed
back their CSI not only to one BS as in current systems, but they broadcast
this information to all collaborating BSs, and the resulting inter-BS CSI
information exchange requirement is minimal.

In the downlink, a major overhead of MCP that needs to be mitigated is
the one of CSI over-the-air feedback (i.e. mobile to base). Furthermore the
collaborating BSs need to exchange the user data to be transmitted through
the backhaul (backhaul overhead). For downlink communication under Frequency Division Duplexing (FDD), each user needs to estimate and feed
back to the system infrastructure (one or more BSs) a number of channel
coefficients, equal at least to the number of collaborating antennas at each
subcarrier in Orthogonal Frequency Division Multiplexing (OFDM). This
feedback load renders the deployment of MCP prohibitive in large scale de-
ployments. In this thesis we suggest the use of a *selective feedback* approach.
In this setup only the significant coefficients are fed back by the users; the
ones whose channel gain exceeds a threshold. This approach can be also
exploited in reducing backhaul overhead through scheduling or precoding
design. It is shown that this is a good tradeoff between performance and
overheads that can facilitate the incorporation of MCP by future systems.

Another promising technique that can increase spectral efficiency of wireless systems is cooperative relaying. In this thesis the utilization of *dynamic
relays* (user terminals relay signals) in cellular systems is investigated. Dynamic relays are more cost effective than static ones, as they bring the
gains of relaying without the need for costly new infrastructure. However
their utilization entails very high overheads and complexities (CSI feedback
requirements, relay selection process). In the present dissertation the performance of dynamic relays in different cellular environments is assessed from
a system level point of view and some novel techniques that exploit dynamic
relays while requiring minimal overhead are presented. The overheads of relaying are proportional to the number of considered relay candidates (relay
selection process). It is suggested that for a specific transmission only a small
but suitable set of relay nodes are considered as relaying candidates. This
is an efficient method to benefit from dynamic relays while circumventing
their drawbacks.

Résumé

La demande sans cesse croissante pour des services sans fil de plus en plus gourmandes en ressources et l'évolution de l'état du marché des communications sans fil, obligent les futurs systèmes (4G) à obéir à des contraintes d'efficacités spectrales plus importantes et à fournir une meilleure qualité de service, particulièrement pour les utilisateurs se trouvant en bordure de cellules. Afin de répondre à ces objectifs, les nouveaux systèmes de communication devront incorporer des technologies qui leur permettent d'augmenter l'efficacité spectrale dans la cellule. Une des techniques des plus prometteuses permettant d'atteindre ces objectifs est le MIMO Coopératif dans les Réseaux Multi-cellulaires (Multicell-MIMO en anglais) qui est capable de diminuer l'interférence intercellulaire et d'augmenter le débit. Cette technique nécessite qu'un certain nombre de stations de base (BSs) se regroupent et traitent les signaux conjointement. Cependant, le Multicell-MIMO coopératif nécessite l'introduction de charges supplémentaires non négligeables sur le réseau afin de permettre le bon fonctionnement de cette technologie. Le but de cette thèse est d'étudier ces charges afin de trouver les mécanismes de les réduire avec des pertes acceptables de performances.

Dans un premier temps, partant du fait que les charges introduites sur le réseau par le Multicell-MIMO coopératif sont proportionnelles au nombre de BSs qui coopèrent, nous sommes concentrés sur le regroupement d'un faible nombre d'entre eux. Nous proposons pour cela une coopération entre les BSs qui génèrent beaucoup d'interférences l'une pour l'autre, au lieu de considérer une coopération entre les stations de bases qui sont géographiquement proches l'une de l'autre. Cette approche nous a permis d'apporter des gains significatifs au niveau de l'efficacité spectrale.

Par ailleurs, la conception centralisée classique pour le MIMO coopératif nécessite que les BSs appartenant à un même groupe soient interconnectées par une unité de contrôle centrale, leur permettant l'échange d'informations entre autres sur l'état du canal (CSI: Channel State Information en anglais). Cette architecture rend difficile le déploiement de cette technologie dans les réseaux de communications mobiles vu qu'elle nécessite des investissements supplémentaires pour le rajout de l'infrastructure additionnelle. Dans cette thèse, on propose une nouvelle approche mettant en œuvre une architecture décentralisée. Ceci nous assure la simplicité d'intégration du principe de la coopération Multi-cellulaire dans les systèmes cellulaires conventionnels à travers de minimes changements sur leur architecture.

En outre, dans les systèmes de communication FDD (Frequency Division Duplexing en anglais) la connaissance du canal de transmission est cruciale afin d'assurer une bonne communication sur le lien descendant. Pour cela, les différents utilisateurs doivent réaliser une estimation du canal et la renvoyer sur une voie de retour. Nous proposons dans cette thèse une approche permettant de diminuer le trafic généré par le processus de retour du CSI en ne sélectionnant que les coefficients du canal ayant un gain supérieur à un niveau préfixé. On montre que la réduction de cette charge sur la voie de retour peut être combinée avec la réduction de la charge sur le backhaul (échange des donnés entre les BSs qui coopèrent).

Une autre technique prometteuse, qui permet d'améliorer l'efficacité spectrale, est l'utilisation des relais. Les relais dynamiques (utilisateurs qui relaient des signaux destinés aux autres utilisateurs) sont plus rentables que les relais statiques, car ils n'exigent pas de déploiement de nouvelle infrastructure coûteuse. Cependant, leur utilisation ajoute des charges supplémentaires et des complexités importantes. Dans cette thèse l'efficacité des relais dynamiques est évaluée dans des environnements différents. De plus quelques techniques originales qui exploitent les relais dynamiques tout en exigeant des charges générales minimales sont présentées.

Résumé des travaux de thèse

Introduction

La demande des services sans fil de haute qualité a connue une augmentation fulgurante au cours des derniéres années. Outre les services traditionnels de la voix, les utilisateurs mobiles demandent d'autres services telles que la navigation Web, la vidéoconférence, le streaming vidéo, la messagerie multimédia, les jeux sur Internet, etc. De ce fait, plusieurs applications nouvelles, qui seront plus en demande dans l'avenir proche, ont été introduites au monde des communications sans fil. Pour faire face à ces transformations, les systèmes cellulaires doivent fournir des débits plus élevés, avoir une couverture omniprésente, et répondre aux exigences de la globalisation et la libéralisation du marché sans fil [1]. Ainsi, les futurs systèmes sans fil devraient [1]:

- Améliorer les débits.

- Avoir une plus grande efficacité spectrale.

- Améliorer la qualité de service des utilisateurs dans les bordures des cellules.

- Être faciles à déployer.

- Être capables de coexister avec les anciennes technologies sans fil.

Afin d'atteindre leurs objectifs, les technologies 4G se sont trouvés obligés de réutiliser des fréquences d'une manière agressive, afin de récolter des gains substantiels sur l'utilisation du spectre. Ces systèmes visent aussi à faciliter la planification cellulaire et le déploiement des Stations de Base (BSs). Cependant ces systèmes soufrent d'importantes pertes de débit provoquées par l'interférence intercellulaire (ICI[2]). Cette perte de performances concerne principalement les utilisateurs situés en bordures des cellules en raison

---

[2]Inter-Cell Interference en anglais

de leur proximité des cellules voisines. L'ICI est donc un facteur qui mène à une dégradation significative des performances et de l'équité du réseau [2]. Toutes ces questions doivent être abordées de manière efficace pour atteindre les objectifs fixés par les spécifications 4G.

Une technique très prometteuse qui permet de diminuer l'interférence intercellulaire et d'augmenter le débit, est le MIMO Coopératif dans les Réseaux Multi-cellulaires (Multicell-MIMO en anglais). Cette technique nécessite qu'un certain nombre de stations de base se regroupent et traitent des signaux conjointement.

MIMO Coopératif dans les Réseaux Multi-cellulaires

Dans cette thèse on propose que dans les systèmes Multicell-MIMO les BSs sont regroupées en groupes de coopération, dont chacun contient un sous-ensemble des BSs du réseau. Le BSs de chaque groupe échangent des informations et traitent conjointement les signaux. Ainsi, les systèmes Multicell-MIMO parviennent à réduire l'ICI et améliorer les performances [3, 4]. Le Multicell-MIMO permet également aux systèmes cellulaires de garder un bon fonctionnement à haut rapport signal sur interférence plus bruit (SINR[3]).

Toutefois Multicell-MIMO impose quelques contraintes qui doivent être étudiées afin de rendre cette technique applicable dans la pratique. Les problèmes introduits par le Multicell-MIMO dans le lien descendant (down-link en anglais) peuvent être classés dans trois catégories [5] (voir la figure suivante):

- *Charges sur la voie de retour*
    - Les utilisateurs estiment plusieurs coefficients de canal (CSI[4]) et les remontent à l'infrastructure du système.

- *Charges sur le réseau backhaul*
    - Les BSs qui coopèrent échangent des données concernant les utilisateurs dans leur zone de couverture.

- Charges sur l'infrastructure du système
    - Unité de Contrôle (UC): le UC rassemble les CSIs, puis elle effectue l'ordonnancement des utilisateurs et conçoit les paramètres de transmission selon la stratégie choisie.

---

[3]Signal-to-Interference-and-Noise Ratio en anglais
[4]Channel State Information en anglais

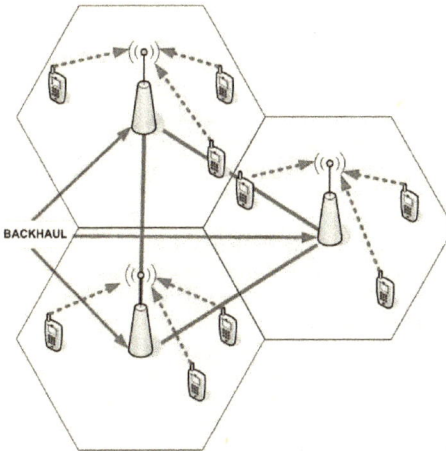

– Liens backhaul à faible latence: Les BSs qui coopèrent sont liées à l'UC via des liens à faible latence pour échanger des CSIs, les décisions sur l'ordonnancement et les paramètres de transmission.

Il faut notez que les charges sur le réseau backhaul sont indépendantes de la conception architecturale du Multicell-MIMO, alors que les charges sur l'infrastructure mentionnées ci-dessus sont liées à la conception classique de l'architecture du Multicell-MIMO [6–9].

Dans la pratique, il est inimaginable que toutes les BSs du système coopèrent [10–12], car toutes les charges susmentionnées sont proportionnels au nombre des BSs qui collaborent, ce qui génère des trafics trop importants.

Certains des algorithmes de regroupement des BSs concernent le problème en lien montant (uplink en anglais) [10, 11, 13]. Il a été montré que le regroupement statique des BSs utilisant du beamforming linéaire améliore significativement l'efficacité spectrale des systèmes cellulaires dans le cas de cellules sectorisées [10, 11]. Ces contributions présentent cependant des limitations dans le cas de gros groupes de BSs vu les charges importantes générées et le manque de diversité créé par l'évolution des conditions des canaux. Un des moyens permettant de diminuer ces charges est de limiter le nombre de BSs par groupe coopératif. Une technique simple qui a été proposée est de permettre la collaboration entre BSs voisins [12]. Toutefois, on montre dans cette thèse qu'à travers des formations dynamiques des groupes coopératifs, on parvient tout de même à récolter des gains significatifs sur la performance sans pour autant trop augmenter les charges [14, 15].

Dans le cadre typique d'un réseau MIMO Multi-cellulaire en FDD[5], chaque utilisateur estime les différents canaux qui le relient aux BSs de son groupe de collaboration (acquisition de CSI). Puis il remonte son CSI à la BS qui lui offre le SNR le plus élevé. Par la suite, la BS transmet ces informations à l'Unité de Contrôle du groupe. L'UC sélectionne les utilisateurs qui vont être servis et calcule les paramètres de transmission pour chaque utilisateur qu'elle transmet ensuite à la BS correspondante. Par conséquent, pour une conception coopérative classique, une UC ainsi que des liens à faible latence entre les BSs et la UC sont nécessaires [6–8]. Ceci impose des changements importants dans l'architecture du système par rapport à ce qui est actuellement déployé et donc génère des coûts de déploiement significatifs. Afin de faciliter la mise en œuvre du Multicell-MIMO, il est primordial de réduire les charges générales qu'implique le déploiement de cette nouvelle conception dans le réseau existant. Cet aspect a été abordé au cours de cette thèse.

Regroupement des Stations de Base

Dans cette section, nous examinons les avantages de la formation de groupes dynamiques de BSs et nous proposons une nouvelle approche dans cette direction. On considère la transmission en lien montant et nous fixons comme objectif la maximisation du débit du système. Il est à noter que la technique proposée peut être généralisée pour s'appliquer au lien descendant. Pour la réception, un beamforming à faible complexité, le Zéro-forcing est utilisé. L'algorithme d'ordonnancement adopté est le round-robin car on veut assurer l'équité dans le système. Cependant, d'autres algorithmes peuvent être

---

[5]Frequency Division Duplexing en anglais

mise en place [11, 12]. L'algorithme de regroupement qu'on propose divise, pour chaque tranche horaire, les BSs disponibles en un ensemble de groupes disjoints. Chaque groupe se voit attribué un sous-ensemble d'utilisateurs présélectionnés pour les servir de manière optimale. Ainsi, chaque groupe forme un réseau distribué d'antennes desservant les utilisateurs qui lui sont rattachés. L'algorithme dynamique pour la formation de groupes est comparé avec des algorithmes statiques et on parvient à montrer que notre approche mène à des gains de performances beaucoup plus importants.

Beamforming Linéaire

Beamforming linéaire a été considéré. La matrice du beamforming est $\mathbf{W}(\mathcal{S}, \mathcal{V}) = \left[\mathbf{w}_1, \mathbf{w}_2, \ldots, \mathbf{w}_{|\mathcal{S}|}\right]^T$. Le signal $\widetilde{y}_i$ reçu par l'utilisateur i est donné par l'expression suivante

$$\widetilde{y}_i = \mathbf{w}_i^T \mathbf{h}_{ii} u_i + \sum_{j \neq i, j \in \mathcal{S}} \mathbf{w}_i^T \mathbf{h}_{ij} u_j + \sum_{k \neq i, k \notin \mathcal{S}} \mathbf{w}_i^T \mathbf{h}_{ik} u_k + \mathbf{w}_i^T n_i \qquad (0.1)$$

où $\mathbf{w}_i \in \mathbb{C}^{\mathcal{V} \times 1}$ est le vecteur du beamforming de l'utilisateur i. Les facteurs $\sum_{j \neq i, j \in \mathcal{S}} \mathbf{w}_i^T \mathbf{h}_{ij} u_j$ et $\sum_{k \neq i, k \notin \mathcal{S}} \mathbf{w}_i^T \mathbf{h}_{ik} u_k$ représentent respectivement l'interférence provenant au sein du groupe et hors groupe coopératif. Le vecteur des signaux que les utilisateurs de l'ensemble $\mathcal{S}$ reçoivent est donné par

$$\widetilde{\mathbf{y}}(\mathcal{S}) = \mathbf{W}(\mathcal{S}, \mathcal{V}) \mathbf{y}(\mathcal{V}) \qquad (0.2)$$

où $\mathbf{y}(\mathcal{V})$ est le vecteur des signaux reçus par les antennes de l'ensemble $\mathcal{V}$. Le SINR de l'utilisateur i est donné par

$$\gamma_i = \frac{\left|\mathbf{w}_i^T \mathbf{h}_{ii}\right|^2}{\displaystyle\sum_{j \neq i, j \in \mathcal{S}} \left|\mathbf{w}_i^T \mathbf{h}_{ij}\right|^2 + \sum_{k \neq i, k \notin \mathcal{S}} \left|\mathbf{w}_i^T \mathbf{h}_{ik}\right|^2 + \left(\left|\mathbf{w}_i^T\right|^2 \sigma^2\right)/p}. \qquad (0.3)$$

La matrice de beamforming est définie comme

$$\mathbf{W}(\mathcal{S}, \mathcal{V}) = \left[\mathbf{H}^H(\mathcal{V}, \mathcal{S}) \mathbf{H}(\mathcal{V}, \mathcal{S})\right]^{-1} \mathbf{H}^H(\mathcal{V}, \mathcal{S}). \qquad (0.4)$$

Il est à noter qu'on peut considérer d'autres techniques linéaires [16]. Avec de Zéro-forcing le SINR de chaque utilisateur i est donné par

$$\gamma_i = \frac{1}{\displaystyle\sum_{k \neq i, k \notin \mathcal{S}} \left|\mathbf{w}_i^T \mathbf{h}_{ik}\right|^2 + \left(\left|\mathbf{w}_i^T\right|^2 \sigma^2\right)/p}. \tag{0.5}$$

Regroupement Statique des Stations de Base

Une solution pratique pour le groupement des BSs serait la formation des groupes qui auraient été pré-spécifiés. Dans ce cas les BSs appartenant à un groupe spécifique ne changent pas au cours du temps [10, 12]. Par conséquent, les groupes sont statiques et les BSs qui ont besoin de coopérer restent toujours les mêmes. Le problème qui se pose dans ce cas est la sélection des régles de groupement statique afin d'optimiser la performance du systéme. Dans cette thèse, on considère que les BSs voisines forment les groupes statiques vu que, ceux sont eux qui, en moyenne, interfèrent les plus les uns avec les autres. Le regroupement statique élimine dans ce cas non seulement une portion des interférences, mais il réduit aussi considérablement les charges générés sur le réseau. Le prix à payer dans cette configuration est que l'interférence n'est pas complétement éliminée et elle limite par conséquent les performances obtenues.

Regroupement Dynamique des Stations de Base

L'objectif du regroupement dynamique des BSs est de former des groupes disjoints (qui correspondent aux graphes disjoints, voir la figure suivante) d'une manière à maximiser la somme-capacité. Le problème de cette maximisation peut être exprimé mathématiquement comme suit

$$C_{max} = \max_{G \in \mathcal{G}} \left\{ R^{(G)} \right\}. \tag{0.6}$$

L'espérance de la somme-capacité du système est donnée par

$$\bar{C} = \mathbb{E}\left[C_{max}\right]. \tag{0.7}$$

Le regroupement statique n'est pas le moyen le plus efficace pour former des groupes des BSs de taille limitée dans un but de coopération car dans ce cas, la macro-diversité n'est pas pleinement exploitée. Un utilisateur pourrait en effet présenter un meilleur canal de transmission avec une BS géographiquement plus éloignée. Par conséquent, pour un utilisateur spécifique, il est plus efficace de recevoir des données à partir des BSs offrant des conditions de canaux les plus favorables indépendamment de leur

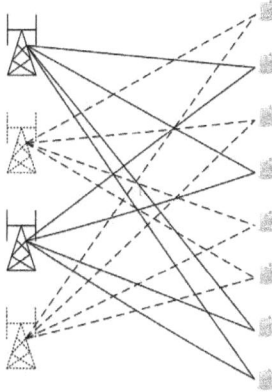

localisations. En outre, les utilisateurs situés en bordure d'un groupe statique sont beaucoup plus sujets à l'interférence intercellulaire et cela compromet l'équité du système. Pour contourner les problèmes susmentionnés, les
groupes de coopération peuvent être formés de manière dynamique. On suppose que chaque groupe coopératif dessert un certain nombre d'utilisateurs
égal au maximum au nombre de ses antennes. On suppose que les utilisateurs
sont associés aux BSs qui leur offrent le SNR le plus élevé. L'algorithme 1,
s'inspiré de [17], a pour but de maximiser la capacité totale du système.

Algorithm 1 Algorithme Avide de Regroupement Dynamique

1: **Étape 1** Déterminer la taille du groupe coopératif.
2: **Étape 2** Commencer par une cellule et ses utilisateurs qui n'ont pas été
   choisis jusqu'ici.
3: **Étape 3.1** Trouver la BS qui maximise la capacité conjointement avec
   la BS qui a déjà été choisie.
4: **Étape 3.2** Répéter l'opération jusqu'á ce que le groupe soit formé.
5: **Étape 4** Répéter les étapes **2**, **3.1** et **3.2** jusqu'à formation de tous les
   groupes coopératifs.

Regroupement Dynamique - Résultats Numériques

On considère un réseau comportant 19 cellules. Les BSs sont situées dans le centre de chaque cellule et chaque BS a une antenne omnidirectionnelle avec un gain de puissance de 9 dB (gain sur la hauteur). Les canaux sont modélisés comme décrit dans la section 2.7. Le coefficient du canal entre l'antenne d'une BS et l'antenne d'un utilisateur est modélisé par (2.19). Le fading multi-trajets est pris en compte qui est décrit par une distribution de Rayleigh (2.6). L'effet de masque suit la distribution log-normale (2.5) avec un écart type de 8 dB et l'atténuation du trajet (pathloss en anglais) correspond au modéle de (2.4).

Dans la première figure qui suit, la somme-capacité des techniques de regroupement est affichée. La somme-capacité par cellule est tracée en fonction du SNR du système (tels que définis dans la section 2.7). Pour une taille de groupement fixée, les BSs non affectés à un groupe forment à eux seuls un groupe plus petit (étant donné qu'on dispose de 19 BSs qui est un nombre premier). A travers ces courbes, on peut voir que les techniques statiques de regroupement apportent bien des gains puisque l'interférence est considérablement réduite. Cependant le regroupement dynamique apporte des gains beaucoup plus significatifs grâce à l'exploitation de la connaissance des CSIs instantanés utilisé pour la formation de groupes. On peut en effet constater qu'un système qui utilise de regroupement dynamique avec une taille de groupement de 2 (2 BSs participent à la coopération) apporte beaucoup plus de gains que les systèmes utilisant des techniques statiques ayant des tailles de groupement beaucoup plus importantes.

Dans la deuxième figure qui suit, la fonction cumulative de la densité de probabilité (CDF[6]) du débit des utilisateurs obtenu par les différents techniques de regroupement est affichée. En plus de l'augmentation de somme-capacité, le regroupement dynamique améliore sensiblement l'équité entre les utilisateurs dans le réseau.

MIMO Coopératif dans un Réseau Decentralisé

Cette thèse vise à étudier les contraintes imposées sur l'infrastructure des systèmes existantes liées à une conception classique pour un système MIMO Multi-cellulaire. Selon cette conception, les BSs de chaque groupe sont interconnectées avec une Unité de Contrôle par liens à faible latence. En supposant un mode de fonctionnement en FDD, un utilisateur envoie son CSI par la voie de retour à une seule BS, généralement celle offrant le meilleur

---

[6]Cumulative Density Function en anglais

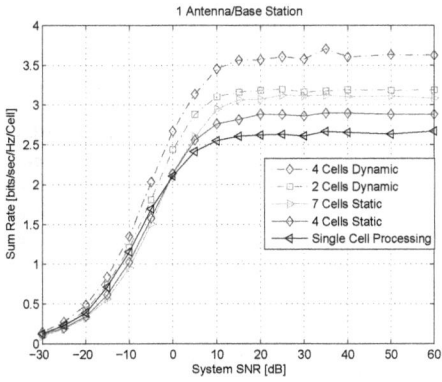

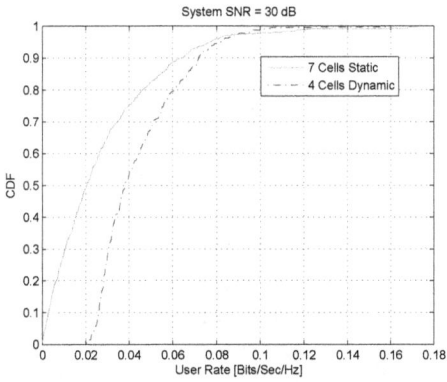

rapport signal sur bruit. Par la suite, chaque BS transmet cette information "CSI local" à l'UC du groupe. Le "CSI local" pour une BS est défini comme le CSI lié aux utilisateurs appartenant à sa cellule. Le "CSI non-local" pour une BS est défini comme le CSI des utilisateurs appartenant aux autres cellules du groupe coopératif. L'UC sélectionne les utilisateurs qui vont être servis (phase d'ordonnancement) et calcule les paramètres de transmission qui sont ensuite transmis à la BS correspondante (phase de transmission). Ainsi, selon la conception classique pour le Multicell-MIMO, une UC et les liens à faible latence sont indispensables [6–8], ce qui exige des changements substantiels au niveau de l'architecture des systèmes actuels et demande par conséquent d'importants coûts. Afin de faciliter le déploiement du Multicell-MIMO, il est nécessaire que les contraintes sur l'infrastructure (UC, liens backhaul à faible latence) soient minimisées.

Feedback Numérique avec d'Erreurs

Dans le cas du feedback numérique, pour chaque utilisateur i il y a un codebook de quantification $\mathcal{C}_i = [\mathbf{c}_1, \mathbf{c}_2, \ldots, \mathbf{c}_N]$ constitué de $N = 2^M$ vecteurs unitaires. Ici M représente le nombre de bits du feedback. Ce codebook est connu aussi bien par l'utilisateur que par l'ordonnanceur. Chaque utilisateur estime[7] son vecteur des coefficients du canal $\mathbf{h}_i$, quantifie sa direction privilégiée $\overline{\mathbf{h}}_i = \mathbf{h}_i / \|\mathbf{h}_i\|$ au vecteur $\mathcal{C}_i$ en se basant sur le codebook. Sa direction privilégiée est l'élément du codebook offrant le plus petit angle avec le quantificateur [18–21]. Par conséquent

$$\widetilde{\mathbf{h}}_i = \mathbf{c}_k, \quad k = \arg \max_{q=1,\ldots,N} \left| \overline{\mathbf{h}}_i^H \mathbf{c}_q \right| = \arg \max_{q=1,\ldots,N} |\cos(\angle(\mathbf{h}_i, \mathbf{c}_q))| \quad (0.8)$$

oú

$$|\cos(\angle(\mathbf{h}_i, \mathbf{c}_q))| = \frac{\left| \mathbf{h}_i^H \mathbf{c}_q \right|}{(\|\mathbf{h}_i\| \|\mathbf{c}_q\|)} \quad (0.9)$$

La qualité de la quantification est déterminée à travers *l'erreur de quantification*, défini comme

$$\sin^2(\angle(\mathbf{h}_i, \mathbf{c}_k)) := 1 - \cos^2(\angle(\mathbf{h}_i, \mathbf{c}_k)). \quad (0.10)$$

Les codebooks doivent être spécifiques pour chaque utilisateur afin d'éviter que plusieurs utilisateurs n'utilisent pas le même vecteur de quantification pour décrire leur direction privilégiée pour le canal de transmission.

---

[7]Dans cette thèse, on suppose une estimation parfaite du canal.

Après la quantification, chaque utilisateur i renvoie par la voie de retour au système l'indice k du vecteur de quantification décrivant au mieux la direction de son canal sous forme binaire. Par conséquent, cette information est définie comme la Direction de l'état du canal (CDI[8]). La taille du codebook détermine l'efficacité de la quantification. En plus du CDI, l'entité qui effectue l'ordonnancement des utilisateurs a besoin de quelques informations concernant la qualité du canal pour pouvoir prendre des décisions de sélection d'utilisateurs. On défini alors le CQI[9] comme indicateur de la qualité du canal. Dans cette thèse, nous considérons la norme du canal non quantifiée $\|\mathbf{h}_i\|$ comme CQI. Cet indicateur ne prend donc pas en compte les interférences entre les utilisateurs. Le choix de cet indicateur simple suffit largement à l'étude des performances du précodage vu qu'on n'est pas intéressé par l'efficacité de l'algorithme d'ordonnancement utilisé.

On considère dans la suite un *codebook aléatoire* puisqu'on n'est pas intéressés par l'optimisation de la conception du codebook. Donc ce codebook contient des vecteurs aléatoires Gaussiens de norme égale à 1, $\mathbf{c}_i \in \sim \mathcal{NC}(\mathbf{0}_B, \mathbf{I}_B)$ and $\|\mathbf{c}_i\| = 1$.

La matrice du canal quantifié est $\widetilde{\mathbf{H}} = \left[\widetilde{\mathbf{h}}_1, \widetilde{\mathbf{h}}_2, \ldots, \widetilde{\mathbf{h}}_K\right]^T$. Les indices binaires qui correspondent aux vecteurs CSI (rangs de $\widetilde{\mathbf{H}}$) ont été remontés à l'entité effectuant l'ordonnancement par des liens sans fils qui introduisent d'erreurs. Par conséquent le CSI possédé par cette éntité est décrit par

$$\widehat{\mathbf{H}} = f\left(\widetilde{\mathbf{H}}\right) \qquad (0.11)$$

où $f(.)$ représente une fonction d'erreurs qui sont introduits par les canaux de la voie de retour.

Lorsque on emploie du feedback numérique, chaque utilisateur i remonte à l'infrastructure du systéme M bits $\mathbf{s}_i^{Tx} = \left[s_{i1}^{Tx}, s_{i2}^{Tx}, \ldots, s_{iM}^{Tx}\right]$, $s_{ij}^{Tx} \in \{0,1\}$ qui correspondent au vecteur $\mathcal{C}_i$ du codebook qui décrit le mieux son CDI. On suppose que chaque bit transmis s_{ij}^{Tx} est reçu erroné avec une probabilité P_e. Donc

$$\Pr\left\{s_{ij}^{Rx} \neq s_{ij}^{Tx}\right\} = P_e. \qquad (0.12)$$

La probilité d'erreur de chaque bit P_e est indépendante et identique dans tous les liens sans fil.

---

[8]Channel Direction Indicator en anglais
[9]Channel Quality Indicator en anglais

Conception Classique Centralisée

Dans cette conception centralisé, chaque utilisateur est associé à une BS qui appartient théoriquement à sa cellule correspondante. Il y a trois phases principales pour une communication descendante dans les systèmes FDD qui prennent en compte le Multicell-MIMO [6–8] (voir les figures suivantes):

1. *Phase 1*

   - Les utilisateurs estiment le CSI lié aux BSs qui coopèrent à l'aide des symboles pilotes. On suppose que l'estimation des canaux est parfaite.
   - Dans le cas où on utilise du feedback numérique, les utilisateurs quantifient la direction de leur canal.

2. *Phase 2*

   - Les utilisateurs remontent leur CSI (numérique ou analogique) par la voie de retour à leur BS principale. Donc chaque BS a le "CSI local".
   - Les BSs transmettent ce "CSI local" à l'Unité de Contrôle du groupe à travers des liens à faible latence. Ainsi l'UC du groupe acquiert de "CSI global".

3. *Phase 3*

   - L'UC effectue l'ordonnancement des utilisateurs basé sur le "CSI global".
   - L'UC détermine les paramètres de transmission et puis les envoie aux BSs.

Le principe de communication ainsi défini a des exigences sur l'infrastructure introduisant des coûts importants pour la mise à niveau des systèmes cellulaires classiques (il faut installer des liens à faible latence entre les BSs et l'UC du groupe). En plus, cela induit une complexification du protocole de communication afin de permettre de bonnes interactions entre les éléments composants du réseau. Toutes ces modifications impliquent inévitablement des changements couteux dans l'architecture actuelle des systèmes cellulaires. Or pour rendre le Multicell-MIMO pratiquement réalisable, il est hautement souhaitable que l'on n'ait pas besoin de faire beaucoup de modifications à la structure existante.

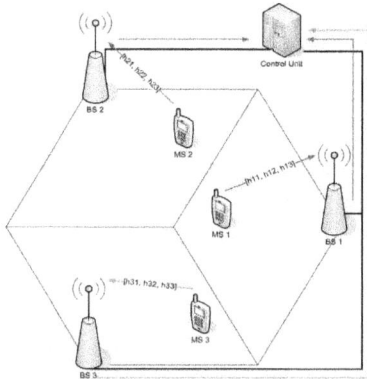

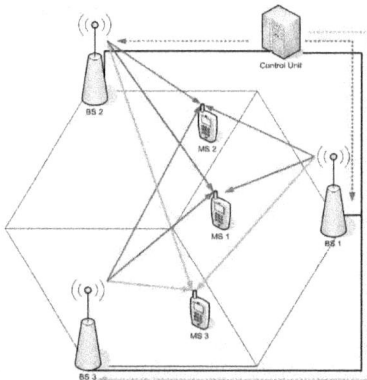

Probabilité d'erreur

Dans le cadre centralisé, chaque utilisateur i utilise juste un lien radio pour remonter les bits $\mathbf{s}_i^{Tx}$ qui decrivent son CSI quantifié, le lien vers sa SB principale. Par conséquent si chaque bit subit une probabilité d'erreur indépendante, la probabilité que les bits $\mathbf{s}_i^{Rx}$ reçus par l'entité qui fait l'ordonnancement est donnée par

$$\Pr\left\{\mathbf{s}_i^{Rx} \neq \mathbf{s}_i^{Tx}\right\} = \sum_{n=1}^{M} \binom{M}{n} P_e^n \left(1 - P_e\right)^{M-n}. \tag{0.13}$$

$\mathbf{s}_i^{Rx}$ est erroné si au moins un de ses bits est erroné.

Conception Decentralisée

Afin d'éviter les désavantages de la structure centralisée, nous proposons une méthode qui ne nécessite pas d'ordonnancement centralisé, mais qui permet toutefois d'atteindre les mêmes performances. Une des raisons à l'introduction du traitement centralisé, est que les BSs impliquées dans un groupe coopératif requièrent la connaissance du "CSI global" alors qu'elles ne disposent que du "CSI local" composé de sous-matrices de $\widehat{\mathbf{H}}$. Le Multicell-MIMO pourrait être réalisé de façon décentralisée si chaque BS impliquée obtenait le "CSI global". Par conséquent les phases sous ces nouvelles contraintes deviennent pour le lien descendant comme suit (voir les figures suivantes):

1. *Phase 1* (identique avec laquelle de la conception centralisée)

   - Les utilisateurs estiment le CSI lié aux BSs qui coopèrent à l'aide des symboles pilotes. On suppose que l'estimation des canaux est parfaite.
   - Dans le cas où on utilise du feedback numérique, les utilisateurs quantifient la direction de leur canal.

2. *Phase 2*

   - Les utilisateurs remontent leur CSI à tous les BSs qui coopèrent de façon omnidirectionnelle. Par conséquent toutes les BSs acquièrent de "CSI global".

3. *Phase 3*

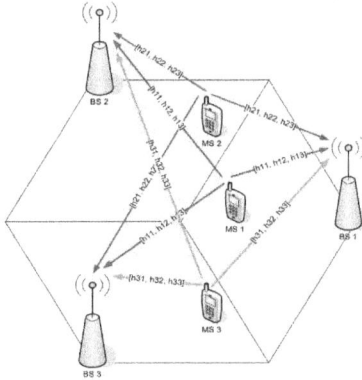

- Les BSs ordonnancent les utilisateurs indépendamment basées sur le "CSI global". Dans le cas où il n'y a pas d'erreurs sur le feedback, les BSs du groupe coopératif choisissent à servir les mêmes utilisateurs.

- Chaque BS conçoit tous les paramètres de transmission et elle garde lesquelles qui correspondent à lui.

Dans ce cas, les coûts en terme d'infrastructure, de signalisation et de complexité des protocoles de communication sont réduits au minimum, puisque les systèmes Multicell-MIMO n'ont plus besoin ni d'UCs ni de liens à faible latence. Par conséquent, l'architecture des systèmes Multicell-MIMO peut rester presque identique aux systèmes cellulaires classiques. Notons que dans ce cadre, les charges sur la voie de retour radio restent cependant les mêmes que celles imposées par une structure centralisée traditionnelle.

Probabilité d'Erreur et Divergence

Dans le cas où des erreurs sont introduites sur la voie de retour, le cas décentralisé peut souffrir plus puisqu'il est plus sensible que le cadre centralisé. En effet, les erreurs introduites sur chaque lien radio utilisé peuvent être différents. Et donc dans le cas centralisé (le cas classique), il n'y a qu'un seul modèle d'erreur affectant les informations transmises puisque chaque utilisateur utilise un seul lien radio pour renvoyer son CSI (CSI qu'il transmet à sa BS principale uniquement). Dans le cadre décentralisé par

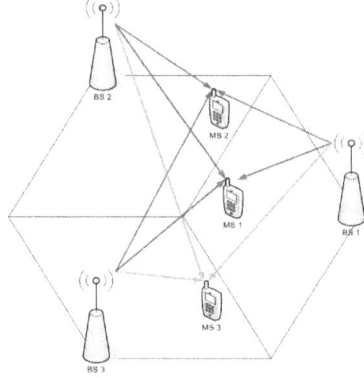

contre, les utilisateurs sont obligés de renvoyer leur CSI à toutes les BSs qui coopèrent, ce qui implique que chaque BS k peut acquérir une version différente du "CSI global" $\hat{\mathbf{H}}_k$.

La probabilité d'erreur de transmission de bits indépendants sur chaque lien de feedback détermine la probabilité que la séquence de bits transmis par chaque utilisateur i, $\mathbf{s}_i^{Tx}$ soit reçue avec des erreurs dans au moins l'un des B BSs qui collaborent. Cette probabilité est donnée par

$$\Pr\left\{\mathbf{s}_i^{Rx} \neq \mathbf{s}_i^{Tx}\right\} = B\left[\sum_{n=1}^{M}\binom{M}{n}P_e^n\left(1-P_e\right)^{M-n}\right]. \qquad (0.14)$$

Les erreurs introduites dans la voie de retour peuvent potentiellement provoquer une dégradation de performances dans le cas décentralisé comparé au cas centralisé. Il est de ce fait primordial d'étudier comment l'existence d'erreurs impacte les performances dans ces deux cas.

On suppose que $\mathcal{B}$ est l'ensemble des BSs qui coopèrent, où $|\mathcal{B}| = B$. A cause des probabilités d'erreur de bits sur chaque lien de retour, quelques BSs peuvent sélectionner des vecteurs de canaux différents pour un même utilisateur. Par exemple, il est possible que le feedback reçu par la BS k, $\mathbf{s}_i^{Rx,k} \neq \mathbf{s}_i^{Rx,\mathcal{J}}$, où $\mathcal{J} \subset \mathcal{B}$, $k \notin \mathcal{J}$. Un ensemble de BSs $\mathcal{J}$, peut recevoir un indice différent du CSI que celui reçu par la BS k. Cette divergence sur la valeur du CSI peut conduire à une dégradation des performances dans le cas décentralisé (dégradation sur l'ordonnancement et sur la conception du précodage).

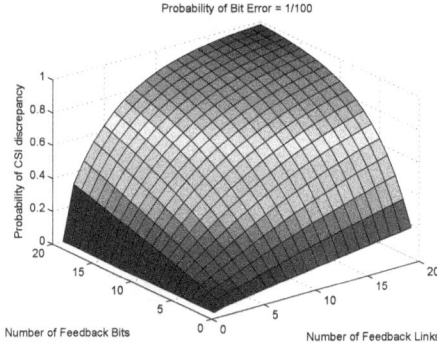

Un bon indicateur de cette divergence des CSIs reçus par les B BSs qui collaborent est la probabilité que cette erreur se produise. Cela constitue aussi un indice traduisant l'écart de performance entre les deux cas de systèmes MIMO Multi-cellulaire. Cette probabilité de divergence du CSI P_d est donnée par l'expression suivante

$$P_d = 1 - \sum_{m=0}^{M} \binom{M}{m} P_e^{B \cdot m} (1 - P_e)^{B \cdot (M-m)}. \qquad (0.15)$$

Le CSI remonté reste identique à travers les B liens que les bits de la même séquence soient correctes ou erronés. Dans ce cas, il n'y a pas de divergence de CSI et la probabilité de cet événement, en considérant B stations de base qui collaborent et que le CSI contient M bits, est donnée par

$$P_{nd} = \sum_{m=0}^{M} \binom{M}{m} P_e^{B \cdot m} (1 - P_e)^{B \cdot (M-m)}. \qquad (0.16)$$

Par conséquent, la probabilité de divergence est $P_d = 1 - P_{nd}$ (0.15). La probabilité de cette divergence en fonction du nombre de bits transmis ainsi que le nombre de liens de transmission indépendants est représentée dans les figures correspondantes. On peut constater que la probabilité d'une divergence CSI s'appuie fortement sur la probabilité d'erreur binaire P_e.

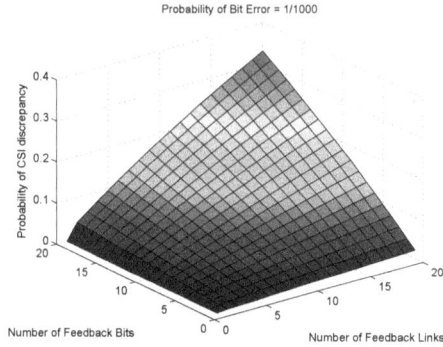

Probability of Bit Error = 1/1000

Conception Decentralisée - Résultats Numériques

Dans cette section, nous évaluons, dans le cas décentralisé qu'on a proposé, la capacité totale du système en fonction des erreurs introduites par la voie de retour et on la compare avec celle du cas centralisé classique. On prend en considération trois secteurs de cellules tri-sectorisées qui interférent mutuellement. On fixe le rayon de la cellule à 1 km vu qu'on étudie un scénario pratique particulier. Les canaux sont modélisés comme décrit dans la section 2.7. Les coefficients du canal entre l'antenne d'une BS et l'antenne d'un utilisateur sont modélisés par (2.19). Le fading multi-trajets est pris en compte qui est décrit par une distribution de Rayleigh (2.6). L'effet de masque suit la distribution log-normale (2.5) avec un écart type de 8 dB et l'atténuation du trajet correspond au modéle de (2.4). Le gain d'antenne en fonction de l'angle horizontal est donné par (2.20). La puissance de transmission est déterminée par le SNR du système, tels que détaillé à la section 2.7.

Nous supposons que chaque utilisateur i obtient une estimation parfaite du vecteur du canal associé à toutes les BSs qui coopèrent ($\mathbf{h}_i$). Dans le cas de feedback numérique, cette estimation est quantifiée et puis transmise de façon omnidirectionnelle (feedback CDI et CQI). Dans le cas centralisé seule la BS principale pour chaque utilisateur reçoit ce feedback (voir la figure correspondante). Dans le cas de la coopération décentralisée toutes les BSs reçoivent le feedback de chaque utilisateur (voir la figure correspondante).

Quand de feedback numérique est employé, le nombre de bits de feed-

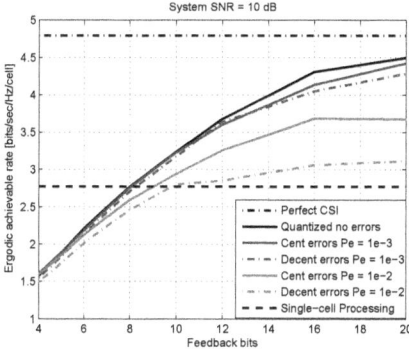

back détermine la performance. Plus précisément, le nombre de bits augmente linéairement avec le SNR et le nombre d'antennes de transmission [20]. Toutefois, la probabilité d'erreur sur le CSI reçu augmente quand le nombre de bits augmente. Ceci peut être vu dans les figures suivantes, où on a affiché la somme-capacité du système en fonction du nombre de bits utilisés avec un ordonnancement type round-robin (le SNR dy système est respectivement de 10 dB et 20 dB, et on utilise un codebook aléatoire). On peut constater que sans l'introduction d'erreurs par la voie de retour et quand le SNR système du est réglé à 10 dB, 16 bits fournissent une bonne approximation du CSI parfaite, alors que pour un SNR du système égal à 20 dB (régime de forte puissance) approximativement 20 bits sont nécessaires.

Les erreurs sur le feedback provoquent inévitablement des dégradations sur les performances de ces deux cas de figure, car certaines informations utiles sont perdues à cause de l'introduction de bits erronés sur la voie de retour. En effet, les erreurs sur les bits du CSI introduisent non seulement des dégradations au niveau de l'ordonnancement des utilisateurs, mais aussi sur la conception du précodage. Le cas décentralisé peut cependant être plus sensible á la dégradation de l'ordonnancement vu que le CSI imparfaite peut causer la sélection d'utilisateurs différents par des stations de base qui coopèrent, dépendamment de l'algorithme d'ordonnancement employé. Ceci va inévitablement augmenter les interférences entre les utilisateurs et dégrader encore plus les performances. Toutefois, l'ordonnancement round-robin est robuste aux erreurs de feedback puisque les décisions de sélection ne sont pas prises en fonction des CSI et cet algorithme d'ordonnancement

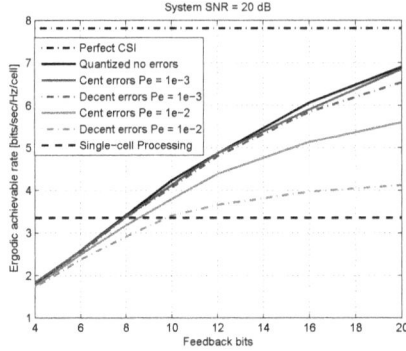

est sélectionné pour la présente évaluation. Il est à noter qu'en absence d'erreurs, les performances des deux cas centralisé et décentralisé coincident avec tout type d'ordonnancement et toute stratégie de transmission.

Dans les figures suivantes la somme-capacité moyenne du système est tracée en fonction de la probabilité d'erreurs binaire respectivement pour un SNR du système de 10 et 20 dB. Lorsque le SNR du systéme a la valeur de 10 dB, 16 bits sont choisis pour les évaluations. Lorsque la SNR système a la valeur de 20 dB, 20 bits sont pris en compte. Cette augmentation de bits est justifiée par la variation de la puissance. En effet plus la puissance est importante plus on a besoin de codebooks grands pour garantir la focalisation des beams et donc de faibles niveaux d'interférence entre les utilisateurs [20]. On peut constater que les erreurs sur le feedback ont par conséquent un impact beaucoup plus grand dans le cas haut SNR (20 bits) qui peut être justifié par (0.13) et (0.14). En outre, dans le cas de 20 bits l'écart de performance entre les deux cas d'étude décentralisé et centralisé augmente. Cette augmentation est justifiée par une plus grande probabilité de divergence du CSI (0.15). Par conséquent, une conception plus intelligente pour le codebook consiste en une réduction du nombre de bits de feedback permettant d'atteindre les mêmes objectifs de performance tout en assurant une meilleure robustesse contre les erreurs.

En règle générale, le cas centralisé est un peu plus robuste aux erreurs de feedback que le cas décentralisé, même si pour une probabilité d'erreur binaire inférieur à 10^{-3}, la différence est négligeable. Il convient également de noter que les résultats présentés ici présentent le cas où aucune technique

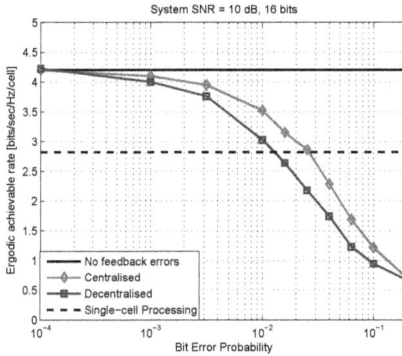

de détection ou de correction d'erreurs n'est employée. Ces techniques peuvent réduire considérablement la probabilité d'erreur binaire P_e, et par la suite peuvent réduire l'impact des erreurs sur le feedback reçu.

Conclusions

Le MIMO Coopératif dans les Réseaux Multi-cellulaires est une technique très prometteuse qui permet à la fois de diminuer l'interférence intercellulaire et d'améliorer les performances. Toutefois le Multicell-MIMO impose quelques contraintes qui doivent être étudiées afin de rendre cette technique applicable. Dans la pratique, seul un nombre limité de BSs peuvent coopérer et traiter conjointement les signaux reçus ou transmis. La solution évidente serait de créer des groupes statiques de taille limitée, même si elle apporte quelques gains de capacité, n'est pas du tout optimale puisqu'elle n'exploite pas pleinement la macro-diversité qui est inhérent aux systèmes multicellulaires. En outre, il compromet l'équité du système puisque les utilisateurs dans la bordure du groupe reçoivent plus d'interférence intercellulaire. Dans cette thèse on a proposé un nouvel algorithme de regroupement dynamique des BSs qui améliore sensiblement les gains de la coopération sans pour autant augmenter fortement les charges générées sur le réseau.

De plus, dans cette thèse, une nouvelle conception architecturale a été proposé qui permet aux systèmes MIMO Multi-cellulaires d'opérer dans un mode décentralisé. En effet, ni une Unité de Contrôle ni les liens á faible latence ne sont nécessaires pour celà. Chaque BS reçoit le CSI de tous les

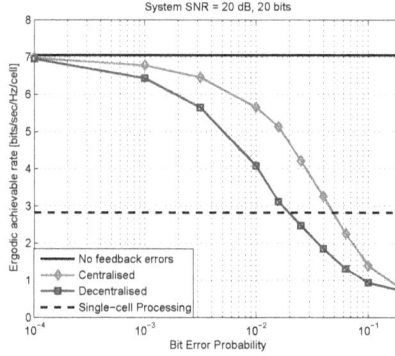

System SNR = 20 dB, 20 bits

utilisateurs ("CSI global") et conçoit la transmission de façon indépendante.
La performance du principe proposé a été évaluée en supposant l'utilisation
d'un feedback numérique et un précodage linéaire (des erreurs sont introduites par la voie de retour). Il a été démontré que la solution proposée
montre peu de dégradation sur la somme-capacité comparée à l'alternative
centralisée. Ces faibles dégradations peuvent être éliminées avec une conception plus intelligente du codebook et/ou avec des techniques de détection
et/ou correction d'erreurs. La conception décentralisée permet ainsi aux
systèmes MIMO Multi-cellulaires d'être déployés avec très peu de modifications sur l'architecture du réseau actuelle et par conséquent avec de faibles
coûts de déploiement.

Contents

List of Figures

Nomenclature

Following there is an outline of acronyms and notations used in this thesis.

Acronyms

Here are the acronyms used in this dissertation which are also defined when they first appear in the text.

1G	First Generation
2G	Second Generation
3G	Third Generation
3GPP	Third Generation Partnership Project
4G	Fourth Generation
ACK	Acknowledge character
AF	Amplify-and-Forward
AMPS	Advanced Mobile Phone Service
AOA	Angle of Arrival
AWGN	Additive White Gaussian Noise
BLER	Block Error Rate
BLR	Backhaul Load Reduction
BS	Base Station
CCI	Co-Channel Interference
CDF	Cumulative Distribution Function
CDI	Channel Direction Indicator
CDMA	Code Division Multiple Access
CoMP	Cooperative Multipoint Transmission
CQI	Channel Quality Indicator
CRC	Cyclic Redundancy Check
CSI	Channel State Information
CSIR	Channel State Information at Receiver

CSIT	Channel State Information at Transmitter
CU	Control Unit
DF	Decode-and-Forward
DL	Downlink
DOA	Direction of Arrival
EDGE	Enhanced Data Rates for GSM Evolution
FD	Full Duplex
FDD	Frequency Division Duplexing
FDMA	Frequency Division Multiple Access
FEC	Forward Error Correction
GPRS	General Packet Radio Service
HARQ	Hybrid Automatic Repeat Request
HD	Half Duplex
HSPA	High Speed Packet Access
ICI	Inter-Cell Interference
IMT	International Mobile Telecommunications
INP	Interference plus Noise Power
ISI	Inter-Symbol Interference
ITU	International Telecommunication Union
LTE	Long Term Evolution
LOS	Line-of-Sight
MAC	Medium Access Control
MCS	Modulation and Coding Scheme
MIMO	Multiple-Input Multiple-Output
MMSE	Minimum Mean Square Error
MRC	Maximum Ratio Combining
MS	Mobile Station
NACK	Negative-acknowledge character
NLOS	No Line-of-Sight
NMT	Nordic Mobile Telephone
OFDM	Orthogonal Frequency Division Multiplexing
OP	Outage Probability
PAPC	Per-Antenna Power Constraint
PDF	Probability Density Function
PFS	Proportionally Fair Scheduling
PHY	Physical Layer
QoS	Quality of Service
RS	Relay Station
RVQ	Random Vector Quantization
SANET	Sensor Actuator Network

SDMA	Space Division Multiple Access
SINR	Signal-to-Interference-and-Noise Ratio
SNR	Signal-to-Noise Ratio
TACS	Total Access Communication System
TDD	Time Division Duplexing
TDMA	Time Division Multiple Access
TTI	Time Transmission Interval
UL	Uplink
ZFBF	Zero-Forcing Beamforming

Notations

Plain letters denote scalars, whereas lower case and upper case boldface symbols denote vectors and matrices respectively, e.g. $\mathbf{y}$ and $\mathbf{Y}$. Unless indicated otherwise, upper case calligraphic letters indicate sets, e.g. $\mathcal{Y}$.

$\mathbb{C}$	The set of complex numbers.
$\mathbb{C}^k$	The complex space with k dimensions.
$\|y\|$	The Euclidean norm of a vector.
$(\mathbf{y})^T$	The transpose operator of a vector.
$(\mathbf{y})^H$	The transpose conjugate operator of a vector.
$\mathcal{CN}$	The complex Gaussian distribution.
$\mathbf{I}_M$	The identity matrix of dimension M.
$\mathbb{E}\left[.\right]$	The expectation operator.
$\mathrm{var}\left[.\right]$	The variance operator.
$|\mathcal{Y}|$	The cardinality of a set.
$\det\left(\mathbf{Y}\right)$	The determinant of a matrix.
$\mathbf{Y}^{\mathrm{T}}$	The matrix transpose.
$\mathbf{Y}^H$	The matrix Hermitian transpose.
$\mathbf{Y}^{-1}$	The matrix inverse.
$[\mathbf{Y}]_{nn}$	The n-th element of matrix's $\mathbf{Y}$ diagonal.
$\mathrm{diag}\left(\mathbf{A}_{11},\ldots,\mathbf{A}_{nn}\right)$	A block-diagonal matrix where $\mathbf{A}_{ii}$ are square blocks.
$\angle\left(\mathbf{x},\mathbf{y}\right)$	The angle between two vectors.
$\exp\left(x\right)$	The exponential function.
$\ln\left(x\right)$	The natural logarithm function.
$\mathrm{erf}\left(x\right)$	The error function.
$\log_2\left(x\right)$	The base 2 logarithm function.
$I_0\left(x\right)$	The modified zero-order Bessel function.

$\boldsymbol{\Gamma}\left(x\right)$	The Gamma function.
$\mathbf{1}_{[m\times n]}$	A matrix with m rows and n columns filled with ones.
$\mathbf{0}_{[m\times n]}$	A matrix with m rows and n columns filled with zeros.
$\odot$	The element-wise multiplication operator.

Chapter 1

Introduction

1.1 Background

The history of cellular networks spans the past three decades. First generation cellular systems (1G) were rolled out for the first time in 1981 (Nordic countries and Saudi Arabia) following the Nordic Mobile Telephone (NMT) standards [22, 23]. The 1G Advanced Mobile Phone Service (AMPS) system was deployed in the USA and worldwide with great success while in the United Kingdom a similar system was developed, called Total Access Communication System (TACS). 1G systems used analog technology, were spectrally inefficient and many of the 1G standards were incompatible[1] [24].

Due to the aforementioned problems plaguing 1G systems, in the late 1980s, European countries consented to a common set of standards forming the second generation (2G) of cellular systems called Global System for Mobile Communications (GSM[2]). GSM uses digital technology and is mainly based on Time Division Multiple Access (TDMA) and also on Frequency Division Multiple Access (FDMA). It supports circuit-switched voice services of a much greater quality than 1G and also text messaging, a feature that proved to be very successful commercially. GSM standards gradu-

---

[1]NMT was employed in the Nordic countries, Switzerland, the Netherlands, eastern Europe, Russia and Saudi Arabia, AMPS was employed in the USA and Australia, TACS in the UK, Radiocom 2000 in France, C-450 in West Germany, Portugal and South Africa, RTMI in Italy whereas various other standards were employed in Japan.

[2]GSM initially stood for Groupe Spécial Mobile.

ally conquered the markets beyond the European continent and became predominant worldwide[3] [23, 24]. As pure GSM supported mainly circuit-switched (voice oriented) services and fairly low data rates, the need for packet-switched oriented services and higher data rates, led to the evolution of GSM. The first GSM enhancement towards more packet-oriented services was called General Packet Radio Service (GPRS), also commercially marked as 2.5G. Enhanced Data Rates for GSM Evolution (EDGE) was a further enhancement of GSM with great significance, marked as 2.75G, providing packet-oriented services.

The great commercial success of GSM initiated discussions on the development of a third generation (3G) of mobile standards. Following this, the International Telecommunication Union (ITU) delivered the International Mobile Telecommunications - 2000 (IMT-2000) standards' framework, defining the key requirements and characteristics that 3G technologies must incorporate. 3G systems were specified to provide packet-oriented services, to be able to furnish greater data rates and carry multimedia traffic. The Third Generation Partnership Project (3GPP), an international collaboration of different parties, was formed with the aim to establish global 3G standards. Universal Mobile Telecommunication System (UMTS) was 3GPP's approved IMT-2000 standard, whose air interface is incompatible with that of GSM as it is based on Code Division Multiple Access (CDMA[4]) [23]. Although the advent of 3G systems was accompanied by great expectations and enormous financial investments (spectrum acquisition costs), 3G technology has been proven less successful commercially than anticipated. This was mainly due to the fact that the enhanced versions of GSM (especially EDGE) provide sufficiently high data rates to support a great range of applications and that pure UMTS did not greatly outperform these GSM variations. Consequently there have been efforts to improve UMTS. An evolution of UMTS has been the High Speed Packet Access (HSPA), commercially marked as 3.5G or 3G+, that provides higher data rates while supporting users of higher mobility. The last evolution effort of UMTS led to the Long Term Evolution (LTE) standard whose release[5] was finalized in March 2009. LTE specifies a fully packet based network capable of meeting the constantly increasing throughput requirements of the market.

---

[3]Other 2G standards are the IS-95 (also known as CDMAOne), IS-136 (also known as D-AMPS) which have been mainly employed in the USA and the PDC employed in Japan.

[4]Contrary to UMTS, the air interface of CDMA2000 which is the 3G evolution of IS-95 in the USA is not incompatible with that of IS-95, as they both rely on CDMA.

[5]LTE release 8

1.1.1 Why 4G?

The demand for high quality wireless services has experienced a profound increase over the past few years. Apart from the traditional voice services, mobile users make use of other applications such as web browsing, videoconferencing, video streaming, multimedia messaging, gaming over the Internet etc. Therefore a whole new spectrum of applications has been introduced to the wireless world which are expected to be in greater demand in the near future. As a result, cellular systems must be able to face the important challenge of meeting the Quality of Service (QoS) demands of each supported application, to provide higher data rates, ubiquitous coverage and to meet the demands of the globalized and liberalized wireless market [1]. Thus it is admitted that future wireless systems should incorporate the following general characteristics [1]:

1. Improved peak data rates.

2. Higher spectral efficiencies.

3. Greater system fairness (improvement of the QoS experienced by the cell-edge users).

4. They are easily deployed.

5. Spectrum flexibility and the ability to coexist with legacy wireless technologies.

In 2008 the ITU published the IMT-Advanced guidelines defining the framework for Fourth-Generation Mobile (4G) standards [25–27]. They define the 4G spectrum, some characteristics that 4G standards must incorporate as well as some minimum requirements that they must meet. 3GPP responded to this with the LTE-Advanced, which has been submitted to ITU in October 2005 and approved as a 4G technology in June 2008 [28–30].

In order to meet their design targets, LTE-Advanced and other 4G candidate technologies, it is necessary to employ *aggressive frequency reuse* (probably full frequency reuse) which not only leads to substantial gains in spectrum usage but also eases cell planning and Base Station (BS) deployment. However aggressive reuse techniques incur important losses in cell throughput resulting from the increased amount of inter-cell interference (ICI). This mainly affects users located on the cell edge as they are much more prone to ICI originating from neighboring cells. Therefore ICI is a factor which leads to significant degradation of performance and fairness

in the network [2]. Furthermore ICI degrades performance of Multiple-Input
Multiple-Output (MIMO) systems, identified as a key 4G feature, hence it
impedes their deployment in the cellular context [31]. All these issues need
to be efficiently addressed in order for the targets set by IMT-Advanced to be
met in practice, leading to a significant improvement of the user experience.

1.2   Thesis Motivation

As bandwidth is scarce and expensive, it is necessary that 4G systems employ efficient *non-bandwidth expanding*[6] interference mitigation techniques
for increasing their spectral efficiency. Spectral efficiency without extra
bandwidth requirements can be mainly achieved by the following technologies:

- Advanced Receiver Processing

- Coordinated Resource Allocation

- Multicell Cooperative Processing (MCP)

- Cooperative Relaying

Advanced receiver processing (interference rejection in the spatial and
other domains) is a complex technique that burdens the receiver, better
suited for the uplink as the BS is an entity capable of baring the cost and
the complexity of the interference rejection algorithms. In the downlink, interference rejection necessarily burdens the Mobile Station (MS) and it substantially increases its complexity and energy consumption [32,33]. Coordinated resource allocation is a method of jointly optimizing the resources of a
wireless system in order to improve performance. In cellular networks significant spectral efficiency gains can be attained by optimizing user scheduling
and power allocation in a multi-cellular context [34–36].

MCP, also called Multicell-MIMO, Distributed Antenna Systems (DAS)
or Cooperative Multipoint Transmission (CoMP) in the LTE standardiza-
tion forum, is an alternative very promising way of achieving spectral efficiencies far beyond the ones that can be attained by coordinated resource
allocation. Compared with advanced receiver processing techniques, MCP
has the advantage that interference mitigation is undertaken by the network
infrastructure in both downlink and uplink; therefore interference mitigation does not burden the MS hardware in the downlink [2, 37–47]. Apart

---

[6]Techniques that mitigate interference without consuming any extra bandwidth.

from MCP, another promising technique for increasing spectral efficiency, robustness, fairness and coverage of wireless systems without consuming extra bandwidth is the utilization of relays [48–54]. Cooperative relaying brings spectral efficiency gains without strictly mitigating ICI. It increases throughput by alleviating the impairments of the wireless channel through transmission path selection.

MCP and cooperative relaying are the two most promising and also most challenging techniques for boosting the performance of future systems and form the core topics of this thesis. Primarily Multicell Cooperative Processing and also cooperative relaying are investigated as they can face many of the technical obstacles impeding the advent of 4G systems [55].

1.2.1   From single-user MIMO to Multicell-MIMO and Cooperative Relaying

The value of multiple collocated antennas (conventional MIMO) has been well appreciated (diversity, array gain, spatial multiplexing, interference suppression) [56, 57]. It has also been admitted that significant gains can be attained when distant antennas cooperate by forming virtual antenna arrays (antenna cooperation). This is very important as it is not always possible to deploy collocated antennas especially on a Mobile Station due to the high cost and the size constraints. Furthermore some scenarios where antenna cooperation is very desirable are intrinsically distributed, e.g. Multicell-MIMO; these scenarios are of a particular interest for this thesis.

In cellular systems we can exploit cooperation of distant antennas in two main ways, namely MCP and cooperative relaying. These two techniques have some inherent similarities; in both cases distant antennas collaborate and have the potential to achieve increased sum-rates and system fairness (especially for the cell edge users) [58–60]. However there are some differences between these two techniques:

- In MCP systems we can engineer the inter-BS links in order to meet specific requirements (i.e. low latency, high capacity). It is possible that optical fibers are utilized for this purpose.

- In relaying enabled networks (cooperative diversity):
  Fixed relay nodes: source to relay links can only be engineered to a certain extent, e.g. Line-of-Sight (LOS) connection can be guaranteed.
  Dynamic relay nodes: source to relay link cannot be engineered, performance cannot be guaranteed.

1.2.2   Multicell Cooperative Processing

In this thesis it is proposed that in MCP enabled systems, BSs are grouped into *cooperation clusters*, each of which contains a subset of the network BSs. The BSs of each cluster exchange information and jointly process signals by forming virtual antenna arrays distributed in space. They can be seen as multiuser MIMO systems where the antennas are no longer collocated but remote. Notably, MCP has been shown to reduce ICI and boost performance [3,4]. MCP also allows cellular systems to operate at higher Signal to Interference plus Noise Ratios (SINRs), and thus enables them to exploit multiple antennas at the BSs which do not provide significant multiplexing gains in low SINR regimes [31].

However MCP comes at the cost of increased feedback, backhaul and infrastructural overheads. In the downlink of cellular systems operating in Frequency Division Duplexing (FDD) mode, the overheads of MCP are related to the inherent need for Channel State Information (CSI) at the transmitter of multiuser MIMO systems [61–65] and also to the distributed nature of collaborative BS processing. The overheads related to MCP can be divided into the following categories [5] (see Fig. 1.1):

- *Feedback Overheads*

    - Users estimate several channel coefficients, equal at least to the total number of cooperating antennas, and feed them back to the system infrastructure.

- *Backhaul Overheads*

    - Cooperating BSs exchange the data aimed at the users in their area of coverage.

- *Infrastructural Overheads*

    - Control Unit (CU): the CU gathers CSI from the BSs, performs scheduling and designs the transmission parameters according to the chosen transmission strategy.

    - Low latency backhaul links: collaborating BSs are connected with the CU via low latency links in order to exchange CSI, scheduling decisions and transmission parameters.

Note that the feedback and the backhaul overheads are independent of the architectural conception for MCP, whereas the infrastructural overheads

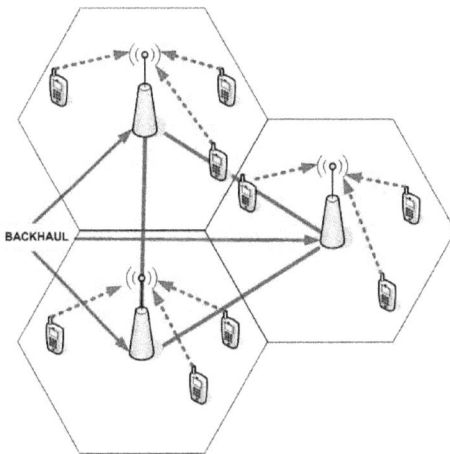

Figure 1.1: The MCP overheads. Red dashed lines represent feedback overhead, green lines represent backhaul overhead and blue lines infrastructural
overheads (CSI exchange).

mentioned above are related to the typical conception for the architecture of MCP [6–9].

In real systems only a limited number of BSs can cooperate in order for the inter-base communication overhead to be affordable [10–12], as all the abovementioned overheads are proportional to the BS cluster size. In [10, 11, 13] some BS clustering algorithms are presented that refer to the uplink problem. Interestingly static clustering of BSs together with linear beamforming significantly improve the spectral efficiency of cellular systems with sectorized cells [10,11]. The limitations in these contributions however are the use of big cluster sizes which yield significant overheads and the lack of diversity with respect to changing channel conditions since cooperation clusters are static. Therefore a natural way for mitigating these overheads is to limit the number of cooperating BSs per cluster. A simple technique that has been proposed is limited static clustering, where BS cooperation groups are of limited size and remain static; only neighboring BSs collaborate and this is a good tradeoff between performance and overhead [12]. However, as it is shown in the present thesis, much higher performance gains can be attained without serious overhead increase if the limited clusters are formed dynamically [14, 15].

According to the typical framework for MCP, in the downlink of FDD systems, a MS estimates the channels related to the BSs of its cooperation cluster. Then it feeds back to the BS of its cell (usually the one that it receives the maximum SNR from) either full or partial CSI (i.e. long-term or quantized CSI). Subsequently, the BS forwards this local information (CSI) to the CU of the cluster which gathers local CSI from all cooperating BSs. Local CSI for a BS is defined as the CSI related to the MSs belonging to its cell. Non-local CSI for a BS is defined as the CSI of the MSs belonging to different cells of the cooperation cluster. The CU selects the users to be served (scheduling phase) and calculates the transmission parameters which are then sent to the corresponding BSs for the transmission to take place (transmission phase). Therefore in the typical conception, a CU and the CU to BSs low latency links are necessary [6–8], a fact which demands substantial changes to the current system architecture and a significant increase in costs. In order to facilitate the implementation of MCP it is necessary that the infrastructural overheads entailed by this conception (CU, low latency backhaul links) are alleviated and this aspect is addressed in the present thesis.

MCP feedback overheads are of major importance as they are proportional to the number of cooperating antennas, which is the sum of the antennas of the clustered BSs. For downlink communication under FDD, users

are required to estimate and feed back all the channel coefficients related
to these antennas in order for the Mulicell-MIMO techniques (multiuser
MIMO techniques where antennas are remote) to be applied. This signaling
load significantly hinders the incorporation of MCP capabilities by future
cellular systems. It is therefore essential that this CSI feedback load is reduced by the application of smart feedback load schemes. Techniques for
reducing CSI feedback load by feeding back a subset of the estimated channel coefficients based on thresholds have been investigated for conventional
cellular networks (where MCP is not enabled) and targeted at multiuser
diversity [66–69]. In MCP enabled networks, feedback load is much greater
as MSs need to feed back the CSI related to all the cooperating BSs and
this comprises a major burden for MCP systems. In the present thesis it
is shown that CSI feedback can be efficiently reduced in MCP enabled networks with the aid of thresholds (selective feedback approach). This can
lead to a good tradeoff between throughput and signaling, permitting MCP
to be brought into practice [70].

In MCP enabled networks each BS potentially transmits to an increased
number of users at a time (space divison multiple access - SDMA) and thus
needs to buffer an increased number of data streams. We will hereby refer
to this as *backhaul overhead*. Consequently MCP enabled systems require
backhaul links of higher capacity which implies an elevated deployment cost.
Therefore reducing backhaul load by routing packets only to BSs that really
need it, is desirable because it can lead to deployment cost reduction. In
addition it can make the use of MCP possible in scenarios where it was
not initially considered feasible due to high bit rate constraints for cell-to-
cell signaling. Techniques for reducing feedback overhead for MCP [10, 12]
and backhaul overhead [45, 47] have been already investigated but without
attempting to jointly minimize both of them. In this dissertation we show
that the feedback reduction provided by the selective feedback approach can
be combined with the reduction of the inter-base backhaul overhead.

1.2.3   Dynamic Relays for Cellular Systems

Apart from MCP, *cooperative relaying* is a promising technique that has the
potential to provide significant spectral efficiency gains. Therefore it is a
technology that can aid cellular networks to transcend their performance
limitations and help them meet the IMT-Advanced requirements.

In the typical dual-hop relay assisted communications, the transmitting
source is aided by one or more relay nodes which together with the source
node form a virtual antenna array. The destination node benefits from

receiving multiple copies of the transmit signal by performing appropriate diversity combining [48,49,51]. The deployment of relaying enabled systems however, is constrained by resource limitations, the need for coordination and increased complexity; therefore their utilization in practice still remains challenging [71].

In cellular systems, for exploiting cooperative diversity, static or dynamic relay stations (RSs) can be considered. The former implies that fixed relay nodes are deployed in specific positions of a cell [52,72–75] whereas the latter implies that MSs act as relays and their position changes in time as users move [76–79]. Furthermore, RS selection and MS scheduling can be performed in a centralized [76,80] or a distributed fashion [53]. Dynamic RS utilization is very cost effective as it does not require extra infrastructural costs and it can attain multi-user diversity gains. However it entails higher complexity and signaling overhead as user mobility renders RS selection rather complicated due to the high CSI feedback requirements. In the present thesis these issues are addressed with the introduction of some novel relay selection techniques which can limit feedback and RS selection overheads in order to exploit dynamic relays in practice.

1.3   Contributions and Thesis Summary

The present thesis focuses primarily on MCP and more specifically on how to limit its overheads in order to bring it into practice. In that respect, several aspects of MCP are investigated, namely BS clustering techniques, feedback reduction techniques and architectural issues. Furthermore a part of the thesis is dedicated to the utilization of dynamic relays (user terminals act as relays) in cellular systems and to some algorithms that render this utilization practically feasible. The contributions of the thesis are partially of conceptual value and partially analytical. We focus on novel feasible schemes and ideas that take practical network constraints into account as much as possible.

Chapter 2 - Propagation over the Wireless Medium

In this introductory chapter there is a brief summary of the fundamental properties and characteristics of wireless channels, namely path-loss, large-scale and multipath fading, frequency selective, space selective and time selective fading. Furthermore this chapter presents all the assumptions made in this thesis regarding wireless channels, the way they are modeled and evaluated.

Chapter 3 - Dynamic Base Station Clustering

In this chapter dynamic ways of creating limited groups of collaborating BSs are investigated. As it has been highlighted above, clusters of cooperating BSs need to be of a limited size in practice as the entailed overhead is proportional to the cluster size (number of BSs participating in each cluster). While in the literature the creation of limited static clusters (each BS always belongs to the same cluster) has been considered, in this chapter we propose that the BS clusters are formed dynamically. In this fashion the BSs that are grouped together are not the ones being geographically closer but rather the ones that interfere the most with each other in a cellular environment plagued by large and small scale fading.

Parts of this chapter's material are contained in:

- A. Papadogiannis, D. Gesbert and E. Hardouin, "A Dynamic Clustering Approach in Wireless Networks with Multi-Cell Cooperative Processing," in Proceedings of *IEEE International Conference on Communications* (ICC 2008), pp. 4033-4037, Beijing, China, May 2008.

Chapter 4 - Decentralized MCP based on Broadcast Feedback

The typical architectural conception for MCP entails that the BSs of each cooperation cluster are inter-connected through a Control Unit playing the role of cluster head. It gathers all the necessary CSI, performs scheduling and coordinates BS transmission. This centralized conception implies an elevated deployment cost for MCP as low latency backhaul links as well as a control unit per cluster are required (*infrastructural overheads*). In this chapter a new decentralized architecture is proposed that circumvents the drawbacks of the typical centralized one while sustaining similar performance levels. According to this, MSs broadcast their CSI to all cooperating BSs (members of the same cluster) and do not only transmit it to the BS providing them the highest Signal to Noise Ratio (SNR). The key problem is the robustness of this scheme which is addressed accordingly.

Parts of this chapter's material are contained in:

- **Patent:** A. Papadogiannis, E. Hardouin, and D. Gesbert, "Multicell Cooperative Communications in a Decentralized Network," International Application No.: PCT/FR2009/051227, Filed: June 2008.

- A. Papadogiannis, E. Hardouin, and D. Gesbert, "Decentralising Multi-Cell Cooperative Processing on the Downlink: a Novel Robust Framework," in *EURASIP Journal on Wireless Communications and Networking*, vol. 2009, Article ID 890685, 10 pages, August 2009.

- A. Papadogiannis, E. Hardouin, and D. Gesbert, "A Framework for Decentralising Multi-Cell Cooperative Processing on the Downlink," in Proceedings of *IEEE Global Communications Conference Workshops* (GLOBECOM 2008), New Orleans, USA, December 2008.

- CELTIC WINNER+ deliverable, "D1.4 Initial Report on Advanced Multiple Antenna Systems," January 2009.

Chapter 5 - Limiting Feedback and Backhaul Overheads of MCP

In this chapter the feedback and backhaul overheads entailed by MCP are addressed. A technique for overhead reduction in the multicell context based on the use of user selective feedback is proposed for mitigating both the feedback and the backhaul overheads. *Selective feedback* has been previously introduced in the context of single-cell processing and targeted at multiuser diversity. In the MCP context, selective feedback relies on users estimating their downlink channel seen from surrounding BSs and deciding, on the basis of a comparison with a pre-determined threshold, whether they should engage in MCP or not. As MCP incurs overhead, the intuition is that only users that will really benefit from it should burden the system with this mode of transmission. We propose an algorithm according to which each MS estimates and feeds back to the system infrastructure the channel coefficients whose average SNR is above an absolute threshold, in order to keep feedback load at prescribed target levels. The multi-cell setting impacts the channel statistics as channels to different BSs undergo different pathloss and large-scale fading (shadowing). The feedback load as a function of a chosen SNR threshold is studied analytically. The other main point made in this chapter is how the feedback reduction can be combined with the reduction of the inter-base backhaul overhead.

Parts of this chapter's material are contained in:

- A. Papadogiannis, H. J. Bang, D. Gesbert, and E. Hardouin, "Selective Feedback Design for Multicell Cooperative Networks," submitted to *IEEE Transactions on Vehicular Technology*.

- A. Papadogiannis, H. J. Bang, D. Gesbert, and E. Hardouin, "Downlink Overhead Reduction for Multi-Cell Cooperative Processing enabled Wireless Networks," in Proceedings of *IEEE International Sym-*

posium on Personal, Indoor and Mobile Radio Communications (PIMRC 2008), Cannes, France, September 2008.

Chapter 6 - Utilizing Dynamic Relays in Cellular Systems

Cooperative relaying is a technique that can admittedly benefit cellular systems to a great extent. Dynamic relays[7] are more cost effective than static ones, as they bring the gains of relaying without the need of costly new infrastructure. However, they incur significant signaling burden to the system (CSI feedback, relay selection complexity). In this chapter the gains from dynamic relaying in different types of cellular environments are evaluated from a system level point of view. Furthermore some techniques are proposed based on thresholds, in order to efficiently exploit dynamic relays while keeping their entailed overheads to a minimum. This is achieved by limiting the number a relay candidates for a specific transmission to a small but suitable set of cell users.

Parts of this chapter's material are contained in:

- **Patent:** A. Papadogiannis, E. Hardouin, and A. Saadani, "Method for reducing the selection complexity of dynamic relays," FR-09 54961, Filed: July 2009.

- A. Papadogiannis, E. Hardouin, A. Saadani, D. Gesbert, and P. Layec, "A Novel Framework for the Utilisation of Dynamic Relays in Cellular Networks", in Proceedings of *IEEE Asilomar Conference on Signals, Systems and Computers* (Asilomar 2008), pp. 975-979, Pacific Grove, USA, October 2008.

- A. Papadogiannis and G. Alexandropoulos, "System Level Performance Evaluation of Dynamic Relays in Cellular Networks over Nakagami-*m* Fading Channels," in Proceedings of *IEEE International Symposium on Personal, Indoor and Mobile Radio Communications* (PIMRC 2009), Tokyo, Japan, September 2009.

- A. Papadogiannis, A. Saadani, and E. Hardouin, "Exploiting Dynamic Relays with Limited Overhead in Cellular Systems," in Proceedings of *IEEE Global Communications Conference Workshops* (GLOBECOM 2009), Honolulu, Hawaii, USA, December 2009.

---

[7]Throughout this thesis we define as dynamic relays, MSs that relay signals intended for other MSs.

Chapter 7 - Other Contributions

The applications of node collaboration certainly exceed the limits of cellular systems[8]. Another group of networks that can significantly benefit from node collaboration is wireless Sensor Actuator Networks (SANETs). In this chapter, inspired by MCP techniques, it is proposed that actuators of a SANET are cooperating (similar to the BS collaboration in cellular systems) in the distribution of scheduling information to sensing nodes. This collaboration exploits macro-diversity and beamforming gains, and can effectively tackle the problem of sensor inactivity (sensors remaining inactive because they cannot correctly decode their scheduling information).

Parts of this chapter's material are contained in:

- M. F. Munir, A. Papadogiannis, and F. Filali, "Cooperative multi-hop wireless sensor-actuator networks: exploiting actuator-cooperation and cross-layer optimizations," in Proceedings of *IEEE Wireless Communications and Networking Conference* (WCNC 2008), Las Vegas, USA, April 2008.

---

[8]Node collaboration in cellular systems is defined as the collaboration between BSs, MSs and/or BSs and MSs.

Chapter 2

Propagation over the
Wireless Medium

The properties of electromagnetic waves heavily depend on their frequency[1]. For wireless communications, *radio waves* and *microwaves* are the suitable bands of the electromagnetic spectrum to be employed. More specifically, the wave propagation models that are discussed in this chapter refer to the UHF[2] and SHF[3] bands (upper frequencies of radio waves and lower frequencies of microwaves) [22,24].

In these bands, transmission of electromagnetic waves over the air is affected by a range of different phenomena, i.e. attenuation, diffraction by obstacles, refraction, reflections and scattering. The intensity of these phenomena highly depends on the frequency/wavelength of the waves. Generally speaking, in lower radio frequencies signals propagate omnidirectionally, signal strength degrades sharply as a function of distance (severe path-loss) and waves permeate obstacles (e.g. building penetration). In higher frequencies waves propagate in straight lines, experience less severe path-loss and bounce-off different objects.

In order to be able to calculate the exact electric field impinging on the

---

[1]Depending on their frequency, electromagnetic waves are divided into Radio waves, Microwaves, Infrared waves, Visible light, Ultraviolet rays, X-rays and Gamma rays.

[2]UHF: Ultra High Frequencies (0.3-3 GHz).

[3]SHF: Super High Frequencies (3-30 GHz).

receive antenna, the Maxwell's equations need to be solved. However solving Maxwell's equations demands very precise information about obstructing objects being available within signal wavelength accuracies. Since it is very hard to obtain information of this sort, some stochastic models have been developed capturing the most important properties of the wireless medium [22, 24]. In this chapter there is a brief description of the models employed in order to approximate the behavior of the wireless medium. Furthermore it is outlined how wireless channels are modeled and evaluated in this thesis.

2.1   Signal Attenuation

In this section the large-scale effects causing signal attenuation are described, namely path-loss and shadowing. These effects determine the average SNR of the received signal, while the instantaneous SNR is affected also by another phenomenon called multipath fading (see section 2.2).

2.1.1   Path-loss

Path-loss (PL) is the effect of power attenuation of an electromagnetic wave caused by the distance and it is frequency dependent. It is defined as

$$PL = \frac{P_r}{P_t} \tag{2.1}$$

where P_t is the transmit power and P_r the received power. The severity of the attenuation due to pathloss depends on the assumed pathloss model. Usually the received power due to pathloss varies over relatively large amounts of space (100-1000m) [24]. Radio propagation over free space experiences the so-called *free space pathloss* given by [4]

$$PL = \frac{\lambda^2 G}{(4\pi d)^2} \tag{2.2}$$

where λ is the signal wavelength, G is the product of the antenna power gains of the transmitter and the receiver and d is the distance between the transmitter and the receiver. Therefore in free space PL is inversely proportional to the square of the distance. This ideal model takes into account only one direct ray and not the effect of reflected ones. The latter is captured by the so-called *ray tracing models* which provide a better approximation of the signal attenuation [22]. As the accuracy of all the aforementioned models is

---

[4]The Friis formula.

limited, empirical models, like the Okomura-Hata and the COST-231, have been extensively employed [24]. A suitable pathloss model for the purposes of the present thesis is the following

$$PL = \beta d^{-\mu} \qquad (2.3)$$

where μ is the *path-loss exponent* and β the *path-loss constant* depending on the average channel attenuation and the antenna characteristics. The path-loss exponent, which determines how rapidly the signal attenuates with distance, in practice varies between 3 and 6. Throughout the present thesis we make use of the 3GPP LTE path-loss evaluation model [81]

$$PL^{dB} = 148.1 + 37.6 \log_{10}\left(d^{km}\right) \qquad (2.4)$$

where the path-loss is in dB and the distance d is in km. This corresponds to $\mu = 3.76$ and $\beta = 10^{-14.81}$.

2.1.2 Large-scale Fading (Shadowing)

Large-scale fading is the effect of random variation of the received signal due to the impediment caused by buildings and large obstacles. Large-scale fading affects the local mean of the received signal and its effect varies over relatively large distances (10-100m) [24]. Shadowing depends on the geographical location of the obstacles and their physical properties as well as the signal wavelength. Shadowing variations are empirically modeled with sufficient accuracy with the aid of log-normal distribution. The probability density function (PDF) of a log-normal distribution is

$$p_\zeta\left(x\right) = \frac{1}{x\sigma_\zeta\sqrt{2\pi}}\exp\left(-\frac{(\ln x - \mu_\zeta)^2}{2\sigma_\zeta^2}\right), \quad x > 0 \qquad (2.5)$$

where μ_ζ and σ_ζ are the logarithmic mean and standard deviation respectively. Throughout this thesis it is assumed that the log-normal distribution has a standard deviation of 8 dB.

2.2 Multipath Fading

Multipath or small-scale fading is the effect of random fluctuation of the received signal due to the constructive and destructive addition of multipath components of the transmit signal. In this section there is a description of the main statistical distributions describing multipath fading.

2.2.1   Rayleigh Fading

The Rayleigh distribution describes the envelope of the small-scale fading process when there is no line-of-sight (NLOS) between the transmitter and the receiver and there is a large number of scatterers (there is no dominant component). This can be inferred from the Central Limit Theorem, as with a sufficiently large number of scattering components, the resulting channel impulse response can be modeled as a zero mean complex-valued Gaussian process $\Gamma \sim \mathcal{CN}(0, \Omega)$. The PDF of its envelope $|\Gamma|$ follows a Rayleigh distribution defined as

$$p_{|\Gamma|}(x) = \frac{2x}{\Omega} \exp\left(-\frac{x^2}{\Omega}\right), \quad x \geq 0 \tag{2.6}$$

where $\Omega = \mathbb{E}\left[|\Gamma|^2\right]$ represents the mean signal strength affected by (2.3) and (2.5). The square of this envelope follows an exponential distribution defined as

$$p_{|\Gamma|^2}(x) = \frac{1}{\Omega} \exp\left(-\frac{x}{\Omega}\right), \quad x \geq 0. \tag{2.7}$$

It has been shown experimentally that Rayleigh fading models with sufficient accuracy Manhattan-type wireless environments.

2.2.2   Ricean Fading

In case there is a LOS component, Rayleigh distribution can no longer accurately describe the fading process. Ricean distribution is a suitable distribution for this purpose, as the severity of fading is described by its $\mathcal{K}$ factor. $\mathcal{K}$ is the power ratio of the direct (LOS) component over the NLOS one.

- For $\mathcal{K} = \infty$ there is only a LOS component, thus there is no multipath fading.

- For $\mathcal{K} = 0$ there is no LOS component and the experienced fading is equivalent to a Rayleigh one.

The PDF of the Rician distribution as a function of factor $\mathcal{K}$ is

$$p_{|\Gamma|}(x) = \frac{2x(\mathcal{K}+1)}{\Omega} \exp\left(-\mathcal{K} - \frac{(\mathcal{K}+1)x^2}{\Omega}\right) I_0\left(2x\sqrt{\frac{\mathcal{K}(\mathcal{K}+1)}{\Omega}}\right), \quad x \geq 0 \tag{2.8}$$

where $\Omega = \mathbb{E}\left[|\Gamma|^2\right]$ and

$$I_0\left(x\right) = \frac{1}{2\pi}\int_0^{2\pi}\exp\left(-x\cos\theta\right)d\theta \tag{2.9}$$

is the modified zero-ordel Bessel function.

2.2.3   Nakagami-m Fading

Rayleigh and Rician fading distributions do not incorporate some physical characteristics of real wireless channels. The Nakagami-m distribution is an empirical and versatile statistical distribution that describes multipath scattering with relatively large delay-time spreads and with different clusters of reflected waves [82]. The PDF of the Nakagami-m distribution is [83, eq. (22)]

$$p_{|\Gamma|}\left(x\right) = \frac{2m^m x^{2m-1}}{\Gamma\left(m\right)\Omega^m}\exp\left(-\frac{mx^2}{\Omega}\right), \quad x \geq 0 \tag{2.10}$$

where $m \geq 1/2$ is the fading parameter, $\Gamma\left(x\right)$ is the Gamma function [84, eq. (8.310/1)], and $\Omega = \mathbb{E}\left[|\Gamma|^2\right]$ is the average fading power. The PDF of the square of this envelope is

$$p_{|\Gamma|^2}\left(x\right) = \left(\frac{m}{\Omega}\right)^m\frac{x^{m-1}}{\Gamma\left(m\right)}\exp\left(-\frac{mx}{\Omega}\right), \quad x \geq 0. \tag{2.11}$$

The Nakagami-m PDF is very general as it can describe other well-known distributions, e.g. for $m = 1$ the Rayleigh and for $m = 0.5$ the one-sided exponential distribution. Moreover, it can be used to model the Rician distribution with sufficient accuracy by setting [85, 86]

$$m = \left[1 - \left(\frac{\mathcal{K}}{\mathcal{K}+1}\right)^2\right]^{-1} \tag{2.12}$$

where $\mathcal{K}$ denotes the Rice factor. Its fading parameter m, can describe the absence or presence of LOS for $m \leq 1$ and $m > 1$ respectively. Moreover, extensive measurement campaigns have shown that the relationship between a signal and its direction of arrival (DOA) can be embodied by m. Hence, varying degrees of fast fading and local scattering can be approximated for any BS-MS and MS-MS channel with the correct choice of m's, leading to accurate modeling of different cellular channel conditions.

Cellular environments can be divided into 3 main categories [87]:

1. *Macrocells*: the cell diameter is usually $0.5 - 20$ km and the antenna
   radiating power is in the order of $0.6 - 40$ W from high towers. LOS
   is usually blocked ($0.5 \leq m \leq 1$) for every BS-MS, MS-MS and ICI
   channel.

2. *Microcells*: the antenna height is a few meters, the radiating power is
   less than 1 W and the cell diameter does not exceed 0.5 km. In such
   systems, there usually exist some BS-MS and/or MS-MS channels with
   $m \geq 1$, while for ICI channels LOS is blocked ($0.5 \leq m \leq 1$).

3. *Femtocells*[5]: they are usually indoor cells whose diameter is $15 - 200$
   m and their antennas radiate some microwatts. In such systems there
   exists some LOS for both useful and ICI channels ($m > 1$) while fading
   of the useful signal can be more severe than the one of the interferers.

In the present thesis we are only interested in Macrocellular and Microcellular environments.

2.3   Frequency Selective Fading

In wireless transmission, different multipath components are reflected and
scattered by obstacles located in close proximity to the transmit/receive antenna (*near scatterers*) or further away from it (*remote scatterers*). The
multipath components resulting from scatterers located close to each other
(usually the near scatterers) arrive at the destination almost simultaneously
and add up either constructively or destructively (multipath fading). The
components resulting from scatterers not in near proximity (often remote
scatterers are not close to each other) arrive at the destination with significant delays. Let the multipath components i and j arrive with delays τ_i and
τ_j respectively. These components are *resolvable* if

$$|\tau_i - \tau_j| >> T_s \tag{2.13}$$

where T_s is the *symbol period*. In this case they constitute separate signal
taps, otherwise they belong to a single tap. An indicative metric of propagation delays is the *delay spread* T_d defined as the propagation time difference
between the first and the last signal tap [88]

$$T_d := \max_{k,n} |\tau_i - \tau_j| \tag{2.14}$$

---

[5]Formerly known as *Picocells*.

for all k, n components. Delay spread is usually proportional to the cell size (larger cells tend to experience greater delay spreads) and causes Inter Symbol Interference (ISI) as delayed replicas of past symbols interfere with current ones. If $T_s >> T_d$ the effect of ISI becomes negligible [89]. Furthermore delay spread gives rise to the phenomenon called *frequency selective fading*, according to which different frequencies of the transmit signal spectrum experience different attenuation levels. A quantity affecting frequency selectivity is the *coherence bandwidth* B_c which is approximately inversely proportional to the delay spread

$$B_c \approx \frac{1}{T_d}. \qquad (2.15)$$

The fading is considered to be *non-frequency selective*, also called *flat fading*, if the signal bandwidth is much smaller than the coherence bandwidth, $B_s << B_c$, or the signal period much greater than the delay spread, $T_s >> T_d$. In this case there is absence of ICI. Else the fading is *frequency selective*.

Broadband channels are generally frequency selective, however throughout this thesis we take into account flat fading channels. 4G systems are envisaged to employ OFDM modulation and it is assumed that the implementation of the MCP methods takes place on a per-subcarrier base, hence the justification of the flat fading channel model.

2.4 Space Selective Fading

The position of the receive (or transmit) antenna influences the phase of each multipath component. Therefore the quality of the received signal, determined by the superposition of these multipath components, fluctuates as a function of the antenna position (space selective fading). An indicative metric of the scale of space selective fading is the distance separating a peak (components add constructively) from a valley (components add destructively), and it is called *coherence distance* D_c [85, 88]. Coherence distance depends on the *angle spread* σ_θ of multipath components around a mean angle of incidence. This spread is introduced by the scatterers; the ones relatively close to the antenna cause severe spread while the more remote ones less severe one.

Coherence distance and angle spread are particularly important for MIMO systems as they determine how correlated is the fading experienced by the different antennas of the array. An upper bound for the angle spread is the *maximum angle separation* $\Delta\theta_{\max}$, $\sigma_\theta \leq \Delta\theta_{\max}$ which is the range where the *power azimuth spectrum* is non-zero. An upper bound for the coherence distance is

$$D_c \leq \frac{\lambda}{2\sin\left(\Delta\theta_{\max}/2\right)}. \tag{2.16}$$

It can be observed that the higher the maximum angle separation is, the shorter the coherence distance becomes.

2.5   Time Selective Fading

In case there is a relative motion of the transmitter, the receiver or an independent change of the scatterers' positions, the phase of each multipath component changes. This may give rise to signal fluctuations varying in time for the duration of these motions. The time-scale of this fluctuation (time to move from a peak to a valley) is called channel *coherence time T_c*. If only the receiver moves with speed u, the coherence time is the time it takes the receiver to cover the coherence distance, $T_c = \frac{D_c}{u}$.

Seen from a different perspective, when there is any kind of motion, each received multipath component k experiences a frequency shift, the so-called *Doppler shift*, $DS_k = f\tau_k$, where f is the signal frequency and τ_k is a function of the motion speed of the k-th component. The maximum difference between the different Doppler shifts is called *Doppler spread D_s* and it is defined as [88]

$$D_s := \max_{k,n} |DS_k - DS_n| \tag{2.17}$$

for all k, n components. The coherence time of the channel is inversely proportional to the Doppler spread

$$T_c \approx \frac{1}{D_s}. \tag{2.18}$$

The channel is considered as *time non-selective* when the duration of each transmit symbol T_s is much shorter than the channel coherence time, $T_s \ll T_c$ (or equivalently that the symbol bandwidth B_s is much larger than the Doppler spread, $B_s \gg D_s$). Else the channel is *time selective*. Throughout this thesis we assume that channels remain time non-selective.

2.6 Fast and Slow Fading

We define a channel as a *fast fading* one when its coherence time T_c is much smaller than the delay requirements of the application. Conversely a channel is characterized as a *slow fading* one if its coherence time is greater than the delay requirement of the considered application [88]. Under fast fading symbol transmission takes place over multiple channel fades. Under slow fading symbol transmission occurs over only one channel fade. In this thesis channels are considered to be slow fading ones.

2.7 Modeling Wireless Channels

In this section it is presented the way that channel coefficients are modeled in this thesis. Furthermore the concept of System SNR is introduced, which the way of calculating the transmit power in all our numerical evaluation scenarios.

Modeling Channel Coefficients

In the present thesis, a generic flat fading channel model that includes antenna power gain, path-loss, large-scale and multipath fading with different fading statistics is considered. Let the network consist of N Base Stations with one antenna each and K single antenna MSs per cell uniformly distributed in the cell area. It is assumed that all BSs communicate on the same frequency (*full frequency reuse*). The channel coefficient between the k-th and the ℓ-th node[6], $k, \ell = 1, 2, \ldots, N + K$, of the network is modeled as

$$h_{k,\ell} = |\Gamma_{k,\ell}| \exp\left(j\,\theta_{k,\ell}\right) \sqrt{G\,\beta\,d_{k,\ell}^{-\mu}\zeta_{k,\ell}} \qquad (2.19)$$

where $|\Gamma_{k,\ell}|$ is the envelope of the multipath fading and $\theta_{k,\ell}$ is the random phase of the channel between the aforementioned nodes that is assumed to be uniformly distributed over the range $[0, 2\pi)$. Multipath fading envelopes follow the Rayleigh (2.6) or the Nakagami-m distribution (2.10). ζ_{ij} is the corresponding log-normal coefficient which models large-scale fading (2.5), $\zeta_{dB} \sim \mathcal{N}(0\,dB, 8\,dB)$. $d_{k,\ell}$ is the distance between the k-th and the ℓ-th node. The path-loss constant β and the path-loss exponent μ are determined from (2.4). G is the product of the power gains of the transmit and the receive antennas. It is assumed that MSs have antennas with unit power

---

[6]A node can be either a BS or a MS.

gain whereas BSs can have either omnidirectional antennas with a 9 dB gain on the elevation or sectorized antennas. In the latter case the sector antenna power gain is a function of the horizontal angle ϕ in degrees as defined in the evaluation parameters of 3GPP LTE [81]

$$G^{dB}(\phi) = 14 - \min\left\{12\left(\frac{\phi}{70}\right)^2, 20\right\}, \quad -180 < \phi < 180. \qquad (2.20)$$

System SNR

The parameter defining the power stemming out of the cell (both for the downlink and the uplink) in each time slot is the System SNR. This is the average SNR received at the edge of the cell when only the serving BS transmits (ICI is not taken into account). Therefore based on this average SNR, the considered path-loss model and antenna gains, the transmit power can be calculated in a straightforward manner.

Chapter 3

Dynamic Base Station Clustering

3.1 Introduction

As it has been outlined in the general introduction, the overheads of MCP are proportional to the number of cooperating BSs (cluster size). Therefore the cluster sizes need to be kept small in order for MCP to become practically feasible. In the literature it has been suggested that neighboring BSs should be grouped together forming limited *static clusters* [10–12]. However static clustering is inefficient as macrodiversity is not fully exploited. In a cellular environment suffering from large and small-scale fading, it is not necessarily the BSs in close proximity that interfere the most as they would in a fading free environment. Consequently, much greater gains can be attained if the limited BS clusters are formed in a dynamic way, exploiting information about fading; in this fashion the BSs that interfere the most can jointly process signals and furnish significant spectral efficiency gains [14,15].

In this chapter we investigate the benefits of dynamic formation of BS cooperation clusters and we propose a new dynamic greedy approach for the formation of the BS clusters. Uplink transmission is considered with the target of sum-rate maximization although the proposed technique can be generalized for the downlink. For the reception Zero-Forcing beamforming (ZFBF) is employed as an example of low complexity MIMO beamforming

scheme. As we are interested in schemes that provide user fairness, the MSs
to be served are selected in a round-robin fashion. However, the algorithm
can be extended for the case of proportionally fair scheduling (PFS) [11,
12]. The BS grouping algorithm divides the available BSs into a number
of disjoint cooperative clusters in each time slot. Each cluster is optimally
assigned to serve a subset of the selected MSs. Thus, each cluster forms a
distributed antenna array which serves the selected MSs associated with it.
The dynamic algorithm for cluster formation is compared with static ways
of forming BS clusters and it is shown that it greatly outperforms them.

3.2   Signal and System Model

The network consists of N base stations with M antennas each and K active single antenna mobile stations overall. An uplink scenario is considered
where a number of B base stations cooperate, where $B \leq N$, and form a
cooperation cluster. Therefore $B \times M$ antennas participate in the cooperation. The antennas of each cluster, under a linear beamforming framework,
jointly combine and process the signal from at most $B \times M$ users simultaneously. The concatenated channel matrix of the system for the uplink within
a cooperation cluster is

$$\mathbf{H} = [\mathbf{h}_1, \mathbf{h}_2, \ldots, \mathbf{h}_{B \times M}]^T \qquad (3.1)$$

where $\mathbf{h}_i \in \mathbb{C}^{K \times 1}$ is the channel vector of the i-th cluster antenna.

Let $\mathcal{B}$ be the set of all disjoint cooperation clusters of $B \times M$ antennas
that are subsets of the overall $N \times M$ antennas of the system. It is assumed
that antennas belonging to the same BS cannot participate in different cooperation clusters. Let $\mathcal{U}$ be the set of all disjoint groups of at most $B \times M$
users that could be possibly scheduled and served by a cooperation cluster
at a time. The proposed system operation scenario is as follows:

- A scheduling algorithm forms a set of cooperation antenna clusters
  $\mathcal{C} \subset \mathcal{B}$, where $|\mathcal{C}| = \frac{N}{B}$ ($|\mathcal{C}|$ needs to be an integer).

- These clusters are mapped to a group of MS clusters $\mathcal{K}$ ($\mathcal{C} \rightarrow \mathcal{K}$),
  where $\mathcal{K} \subset \mathcal{U}$ and $|\mathcal{K}| = |\mathcal{C}|$.

Let $\mathcal{V} \in \mathcal{C}$ be one of the selected antenna clusters and $\mathcal{S} \in \mathcal{K}$ the MS
cluster mapped to it ($\mathcal{V} \rightarrow \mathcal{S}$) by the scheduler. Thus $\mathcal{S}(\mathcal{V})$ is the MS cluster which will be served by the $\mathcal{V}$ group of cooperating antennas. Therefore
$\mathbf{H}(\mathcal{V}, \mathcal{S})$ is the uplink channel matrix related to this BS cluster and group

of MSs, $\mathbf{y}$ is the received signal vector by the BS antennas, $\mathbf{u}$ is the vector of transmit symbols and $\mathbf{n}$ is a vector of independent complex circularly symmetric additive white Gaussian noise (AWGN) components, $n \sim \mathcal{NC}\left(0, \sigma^2\right)$. Therefore $\mathbb{E}\left[\mathbf{n}\mathbf{n}^H\right] = \sigma^2 \mathbf{I}_{B \times M}$. Let $\mathbf{u} = \left[u_1, \ldots, u_{|\mathcal{S}|}\right]^T$ be the vector of transmit symbols with power $\mathbf{p} = \left[p_1, \ldots, p_{|\mathcal{S}|}\right]$ where $p_i = \mathbb{E}\left[|u_i|^2\right]$. Equal power allocation across MSs is assumed, $\mathbf{p} = p\mathbf{1}_{[|\mathcal{S}| \times 1]}$, where $\mathbf{1}$ is a vector of 1s with dimension $|\mathcal{S}| \times 1$.

3.2.1 Linear Beamforming Model

Linear beamforming has been considered for its low complexity and the beamforming matrix is $\mathbf{W}\left(\mathcal{S}, \mathcal{V}\right) = \left[\mathbf{w}_1, \mathbf{w}_2, \ldots, \mathbf{w}_{|\mathcal{S}|}\right]^T$. The extracted signal $\widetilde{y}_i$ of MS i is

$$\widetilde{y}_i = \mathbf{w}_i^T \mathbf{h}_{ii} u_i + \sum_{j \neq i, j \in \mathcal{S}} \mathbf{w}_i^T \mathbf{h}_{ij} u_j + \sum_{k \neq i, k \notin \mathcal{S}} \mathbf{w}_i^T \mathbf{h}_{ik} u_k + \mathbf{w}_i^T n_i \qquad (3.2)$$

where $\mathbf{w}_i \in \mathbb{C}^{\mathcal{V} \times 1}$ is the beamforming vector related with MS i. The factors $\sum_{j \neq i, j \in \mathcal{S}} \mathbf{w}_i^T \mathbf{h}_{ij} u_j$ and $\sum_{k \neq i, k \notin \mathcal{S}} \mathbf{w}_i^T \mathbf{h}_{ik} u_k$ correspond to intra-cluster and inter-cluster intererence respectively. The term $\mathbf{w}_i^T n_i$ corresponds to the *noise enhancement* effect. In matrix notation the extracted signal for the $\mathcal{S}$ users is

$$\widetilde{\mathbf{y}}\left(\mathcal{S}\right) = \mathbf{W}\left(\mathcal{S}, \mathcal{V}\right) \mathbf{y}\left(\mathcal{V}\right) \qquad (3.3)$$

where $\mathbf{y}\left(\mathcal{V}\right)$ corresponds to the received signal of the cooperating antennas $\mathcal{V}$. The SINR of MS i is

$$\gamma_i = \frac{\left|\mathbf{w}_i^T \mathbf{h}_{ii}\right|^2}{\displaystyle\sum_{j \neq i, j \in \mathcal{S}} \left|\mathbf{w}_i^T \mathbf{h}_{ij}\right|^2 + \sum_{k \neq i, k \notin \mathcal{S}} \left|\mathbf{w}_i^T \mathbf{h}_{ik}\right|^2 + \left(\left|\mathbf{w}_i^T\right|^2 \sigma^2\right)/p}. \qquad (3.4)$$

The beamforming matrix is chosen to meet the Zero-Forcing criteria as follows

$$\mathbf{W}\left(\mathcal{S}, \mathcal{V}\right) \mathbf{H}\left(\mathcal{V}, \mathcal{S}\right) = \mathbf{I}_{|\mathcal{S}|} \qquad (3.5)$$

where $\mathbf{I}_{|\mathcal{S}|}$ is an identity matrix with the dimension equal to the number of selected users $|\mathcal{S}|$. Therefore the Moore-Penrose pseudoinverse of the channel is selected as the beamforming matrix

$$\mathbf{W}\left(\mathcal{S},\mathcal{V}\right) = \left[\mathbf{H}^H\left(\mathcal{V},\mathcal{S}\right)\mathbf{H}\left(\mathcal{V},\mathcal{S}\right)\right]^{-1}\mathbf{H}^H\left(\mathcal{V},\mathcal{S}\right). \qquad (3.6)$$

Note that other choices of receiver processing can be considered (e.g. see [16]). With ZFBF intra-cluster interference is eliminated and the SINR is

$$\gamma_i = \frac{1}{\displaystyle\sum_{k\neq i,k\notin\mathcal{S}} \left|\mathbf{w}_i^T\mathbf{h}_{ik}\right|^2 + \left(\left|\mathbf{w}_i^T\right|^2\sigma^2\right)/p}. \qquad (3.7)$$

3.2.2   Graph Interpretation

The problem of the formation of the clusters of BSs that will serve the MSs can be expressed by the aid of graphs. Sum-rate is targeted to be maximized. The constraint is that the graphs formed by connecting BSs (which form clusters) and MSs need to be disjoint, since each BS and MS can belong to a single BS and MS cluster respectively.

Let $\mathcal{G} = \{G = [V, E]\}$ be the constrained graph set where BSs are arranged into disjoint clusters and each cluster is connected to an MS set such that all MS sets are disjoint. V stands for the *vertices* and E stands for the *edges* of the graph. In this case the vertices are the BSs and the MSs while the edges are the connections between them. The evaluation metric is the system sum-rate which is given by the following expression

$$R^{(G)} = \sum_{\mathcal{V}\in G}\sum_{k\in\mathcal{S}(\mathcal{V})} \log_2\left(1+\gamma_k\right). \qquad (3.8)$$

As an example, the case of 4 BSs with 2 antennas each is shown in Fig. 3.1. The cluster size is 2, which implies that each cluster consists of 2 BSs. Since each cluster has 4 antennas, it can serve up to 4 MSs simultaneously in a spatially orthogonal way.

3.2.3   Static BS Clustering

A practically feasible solution for MCP would be the formation of some pre-specified BS clusters. In this case BSs that form each specific cluster do not change in time [10, 12]. Therefore clusters are static and BSs that need to cooperate remain the same. The problem arising in this case is the selection of the BSs that shall form the static clusters in order for the system performance to be maximized. In this chapter neighboring BSs are chosen to form the static clusters, as they are the ones that on average interfere the most with each other in a conventional cellular system. Static

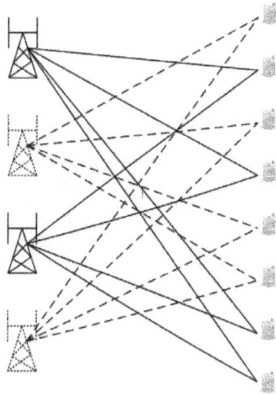

Figure 3.1: A graph representation of the case of 4 BSs with 2 antennas each.

clustering eliminates only a fraction of the inter-cluster interference but it dramatically reduces the inter-base communication burden of the case entailing full network cooperation. The cost is that inter-cluster interference is not completely eliminated and therefore system performance remains ICI limited. Furthermore the MSs at the edge of the cluster suffer more from inter-cluster ICI and therefore system fairness is compromised.

3.3 Dynamic Clustering Based Coordination

In this section there is a description of some cooperative schemes that aim to maximize the sum-rate of the system. The target is to form the disjoint graphs in a way that maximizes the sum-capacity. The problem of sum-capacity maximization can be expressed mathematically as follows

$$C_{max} = \max_{G \in \mathcal{G}} \left\{ R^{(G)} \right\}. \tag{3.9}$$

The expected value of the achievable sum-rate of the system is

$$\bar{C} = \mathbb{E}\left[C_{max}\right] \tag{3.10}$$

where $\mathbb{E}[.]$ is the expectation operator over all channel realizations and MS locations.

3.3.1   Full Coordination

It is assumed that MSs are scheduled in a round-robin fashion in order to provide fairness. In each time slot a number of MSs equal to the total number of antennas in the system is selected. The optimal MCP strategy in a cellular network would require that all BSs be inter-connected and form a single cooperation cluster. The BSs perform joint beamforming and serve the selected users simultaneously by forming a large distributed antenna array. The signal extraction can take place in a Control Unit which gathers all the CSIR of the network and designs the beamforming matrix. This scheme can ideally eliminate inter-cluster interference and bring enormous sum-rate gains [2,40]. However such a scheme is practically infeasible due to the extremely high inter-base communication requirements; all CSI of the network and user signals need to be routed to the CU.

3.3.2   Greedy Dynamic Multicell Processing

Static MCP is not the most efficient way of forming cooperation clusters of limited size. This is because by forcing specific BSs to cooperate, the macro-diversity provided by the distributed nature of MCP is not fully exploited. A MS might experience much better channel conditions to a more distant BS than to a closer one due to the randomness of small and large-scale fading. Therefore for a specific MS it is more effective to receive useful data from the BSs providing the most favorable channel conditions irrespective of their geographical location. In addition, MSs located at the edge of a static cluster are much more prone to ICI originating from neighboring clusters than the ones at the centre of the cluster. This compromises system fairness since MSs at the cluster border will always experience a degraded performance.

To circumvent the aforementioned problems, cooperation clusters can be formed dynamically. It is assumed that each cooperation cluster serves a number of MSs equal to its number of antennas. Due to round-robin scheduling, specific MSs need to be served at each cell at a time. It is assumed that MSs are associated with the BSs that they receive the strongest SNR from. Algorithm 2, inspired from [17], is proposed for sum-capacity maximization with adaptive MCP.

By introducing intelligence in the way that the BSs form clusters in order to serve the selected MSs, the sum-rate increases significantly together with

---

Algorithm 2 Greedy Dynamic Base Station Clustering

1: **Step 1** Specify the cluster size (number of cooperating BSs).

2: **Step 2** Start from a random cell that has not been chosen so far. This corresponds to one BS and some specific MSs, assigned to this BS, that need to be served in the present time slot.

3: **Step 3.1** Find the BS (with the MSs associated with it) that maximize the joint capacity with the initial BS and MSs.

4: **Step 3.2** Continue in the same fashion until the BS cluster is formed (the specified cluster size is reached). B bases and $B \times M$ users are connected.

5: **Step 4** Go to step 2 until all the BS clusters are formed.

---

fairness across users. This is because clusters change dynamically, and there are no cluster regions constantly at the edge and always very prone to ICI.

The greedy clustering algorithm benefits more clusters formed earlier than the ones formed at a later stage as there are fewer choices of BSs available for selection to clusters formed later. In order for this fairness issue to be overcome, each clustering formation phase starts from a random cell and not from a specific one (step 2). Therefore, on average, there are no BSs favored more than others.

A CU is needed in order to gather the CSIR and run the adaptive algorithm for cluster formation. The fact that BS clusters are formed dynamically means that in each time slot different antennas perform coherent combining of the signals in order to serve the MSs. The signal extraction can take place at distributed CUs (one per cluster), a fact which implies that the received signals need to be routed to the cluster CU. Therefore routing burden and inter-base communication requirements of the full coordination case (section 3.3.1) are dramatically reduced.

3.4 Numerical Evaluation

A network consisting of two tiers of cells has been considered ($N = 19$ cells overall). BSs are located in the centre of each cell and each BS has one omnidirectional antenna ($M = 1$) with a power gain of 9 dB (gain on the elevation). Channels are modeled as described in section 2.7. The channel coefficient between the i-th antenna and the j-th MS is given by (2.19). NLOS channels are assumed, thus multipath fading is described by the Rayleigh distribution of (2.6). Shadowing follows the log-normal

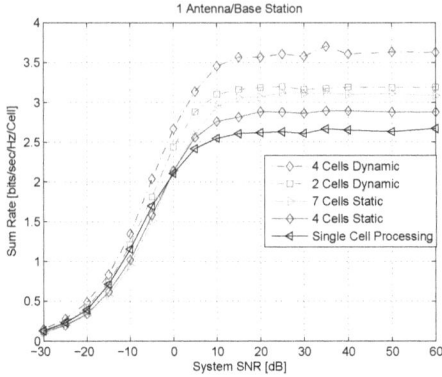

Figure 3.2: Sum-rate per cell versus the system SNR (uplink) for different cluster sizes.

distribution of (2.5) with standard deviation 8 dB and path-loss follows the model of (2.4).

In Fig. 3.2 the average sum-rate performance of the different clustering techniques is shown. The average sum-rate per cell is plotted against the system SNR (as defined in section 2.7). For a specific cluster size, when all clusters have been formed, the remaining BSs (less than the cluster size) form a smaller cluster. This is since there are 19 BSs overall which is a prime number. It can be seen that static clustering MCP techniques outperform single cell processing as the amount of ICI is significantly reduced. The proposed dynamic clustering scheme provides significant sum-rate gains as it exploits the knowledge of instantaneous CSI in the formation of clusters. A dynamic clustering scheme with cluster size of 2 (2 BSs participate in the cooperation) outperforms static clustering schemes with much larger cluster sizes.

In Fig. 3.3 the cumulative distribution function (CDF) of the user rates for two different clustering schemes can be seen. Except from sum-rate increase, dynamic clustering significantly improves fairness amongst the MSs of the network. This can be seen by the fact that the CDF of the dynamic grouping scheme is steeper than the one corresponding to static grouping.

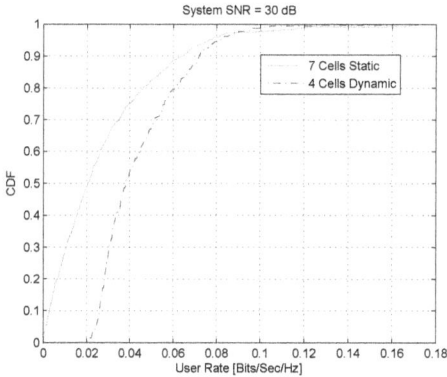

Figure 3.3: CDF of the user rates (100 users/cell) for a static cluster of 7 cells and a dynamic cluster of 4 cells.

3.5   Conclusions

MCP has been proposed as an effective way of facing ICI and increasing spectral efficiency in cellular systems. Its main drawback is the necessity of increased signaling and inter-base communication. In practice, only a limited number of BSs can cooperate and jointly process the received or transmit signals, in order for these overheads to be affordable. The obvious solution of creating static limited clusters of cooperating BSs, even though it provides some capacity gains, is not optimal as it does not fully exploit the macro-diversity which is inherent to MCP. Furthermore it compromises system fairness since users at the cluster edge experience degraded performance, as they are more prone to ICI. In this chapter a novel algorithm has been proposed for dynamic BS clustering which leverages the knowledge of the instantaneous channel state. It groups BSs in a dynamic way that maximizes the sum-rate performance of the MSs to be served in each time slot. It has been shown that this strategy leads to significant sum-rate gains and enhances the fairness of the system comparing to static clustering schemes while keeping the cluster sizes small.

Chapter 4

Decentralized MCP based on Broadcast Feedback

4.1 Introduction

In the previous chapter one major issue related with the overhead of MCP has been addressed, the efficient reduction of BS cluster sizes. The cluster size impacts all the different types of overhead related with MCP and thus the smaller each cluster becomes, the easier it is for MCP to be deployed. Although efficiently limiting cluster sizes can relieve future wireless systems from a great burden, it is not sufficient to render MCP practically feasible. For deploying MCP we need BS clusters of a limited size as well as low signaling and infrastructural overheads.

This chapter aims to address the infrastructural overheads related with the typical conception of MCP. According to this, the BSs of each cluster are inter-connected with a CU through low latency backhaul links. Assuming an FDD framework of operation, a MS feeds back to only one BS (usually the one providing the best SNR) the CSI related to the BSs of the cooperation cluster where it belongs. Subsequently, each BS forwards this local information (CSI) to the CU of the cluster which gathers local CSI from all cooperating BSs. Local CSI for a BS is defined as the CSI related to the MSs belonging to its cell. Non-local CSI for a BS is defined as the CSI of the MSs belonging to different cells of the cooperation cluster. The CU selects the users to be served (scheduling phase) and calculates the transmission parameters which are then sent to the corresponding BSs for the transmission to take place (transmission phase). Therefore according to the typical MCP conception, a CU and the CU to BSs low latency links are necessary [6–8], a fact which demands substantial changes to the current system architecture and a significant increase in costs. In order to facilitate the deployment of MCP it is necessary that the infrastructural overheads entailed by this conception (CU, low latency backhaul links) are alleviated.

In this chapter a framework for decentralizing MCP is introduced which aims at keeping the necessary infrastructural overheads and costs for accommodating MCP to a minimum. We propose a new feedback architecture according to which each BS collects local together with non-local CSI; each MS broadcasts its CSI estimate to all cooperating BSs. In this case, each BS can perform scheduling and design the transmission parameters independently, without the need for any CSI exchange with a central entity; the same scheduling decisions are made by each BS. The proposed framework has a potential sensitivity to feedback errors since MSs utilize several radio links in order to communicate their CSI to the collaborating BSs. This sensitivity is evaluated in the present chapter under a practical linear precoding framework [90]. It is shown that the proposed decentralized framework is robust against feedback errors and can effectively circumvent the drawbacks of the typical MCP framework, facilitating the advent of MCP in real systems.

4.2   System Model

A cellular system is considered which comprises B base stations and K mobile stations overall. We consider the case of single antenna BSs and MSs for simplicity, although our results can be easily generalized to the multiple antenna case. Downlink communication is taken into account and frequency flat fading is assumed. The received signal of the i-th MS can be described as

$$y_i = \mathbf{h}_i^T \mathbf{x} + n_i \tag{4.1}$$

where $\mathbf{h}_i = [h_{i1}, h_{i2}, \ldots, h_{iB}]^T$ represents the channel vector of the i-th user, $\mathbf{x} \in \mathbb{C}^B$ is the vector containing the transmit antenna symbols and $n_i \sim \mathcal{NC}\left(0, \sigma^2\right)$ is the independent complex circularly symmetric AWGN coefficient.

Assumptions

Average per antenna power constraints (PAPCs) have been considered,

$$\mathbb{E}\left[|x_n|^2\right] \leq P_n, \quad n = 1, \ldots, B. \tag{4.2}$$

It is assumed that the system operates in FDD mode and that each MS i obtains a perfect estimate of its own channel state $\mathbf{h}_i$. In addition, we consider delayless feedback links which are utilized by the MSs in order to feed back their CSI to the system infrastructure. The fed back CSI by the users can be corrupted corrupted by errors or noise (depending on the feedback framework) introduced by the feedback channel.

Single-Cell Processing

In the case of single-cell processing (absence of BS cooperation), each MS receives useful signal only from one BS, usually the one providing the best channel gain. Since single antenna BSs are assumed, in each time slot B MSs are scheduled for transmission. The vector containing the transmit symbols $\mathbf{u} = [u_1, \ldots, u_B]^T$ is mapped directly to the transmit antennas $\mathbf{x} = \mathbf{u}$. Therefore the i-th MS receives the following signal when k is its associated BS

$$y_i = h_{ik}u_k + \sum_{j=1, j\neq k}^{B} h_{ij}u_j + n_i \tag{4.3}$$

where h_{ik} corresponds to the channel coefficient related to the useful signal and $\sum_{j\neq k} h_{ij}u_j$ corresponds to the detrimental ICI. Thus the SINR of the user i is

$$\gamma_i = \frac{|h_{ik}|^2 p_k}{\displaystyle\sum_{j=1, j\neq k}^{B} |h_{ij}|^2 p_j + \sigma^2} \tag{4.4}$$

where $p_k = \mathbb{E}\left[|u_k|^2\right]$ and $p_j = \mathbb{E}\left[|u_j|^2\right]$ represent the respective power allocation levels. In this chapter equal power allocation is considered across MSs.

4.3   Linear Precoding for MCP

In MCP enabled networks each group of collaborating BSs forms a distributed antenna array. Therefore all the typical multi-user MIMO precoding techniques can be applied in order for the ICI to be mitigated. In this chapter linear precoding is considered for MCP transmission as it provides a good tradeoff between performance and complexity and it is also more robust to imperfect CSI compared to non-linear schemes [18, 19, 64, 65]. Furthermore linear precoding together with the more practical quantized feedback can be optimal under certain circumstances [91]. In addition to this, linear precoding scales optimally when a large number of MSs is available and opportunistic scheduling is employed [90].

Thus if B single antenna BSs jointly perform linear precoding on the downlink, the BS antennas combine and serve at most B single antenna mobile stations simultaneously. The complete channel matrix of the system is

$$\mathbf{H} = [\mathbf{h}_1, \mathbf{h}_2, \ldots, \mathbf{h}_K]^T \qquad (4.5)$$

where $\mathbf{h}_i \in \mathbb{C}^B$ is the channel vector of the i-th MS. Let $\mathcal{S}$ be the set of MSs scheduled to be served in a specific time slot, where $|\mathcal{S}| \leq B$. Therefore $\mathbf{H}(\mathcal{S}) = [\mathbf{h}_1, \mathbf{h}_2, \ldots, \mathbf{h}_{|\mathcal{S}|}]^T$ is the channel matrix related to these MSs. The vector of transmit symbols $\mathbf{u} = [u_1, \ldots, u_{|\mathcal{S}|}]^T$ with power $\mathbf{p} = [p_1, \ldots, p_{|\mathcal{S}|}]^T$, where $p_i = \mathbb{E}\left[|u_i|^2\right]$, is mapped to the transmit antennas as follows

$$\mathbf{x} = \mathbf{W}\mathbf{u}. \qquad (4.6)$$

$\mathbf{W} = [\mathbf{w}_1, \mathbf{w}_2, \ldots, \mathbf{w}_{|\mathcal{S}|}]$ is the precoding matrix of size $B \times |\mathcal{S}|$ which is a function of the received CSI of the scheduled users and $\mathbf{w}_i \in \mathbb{C}^B$ is the beamforming vector corresponding to MS i. Therefore with linear precoding, the i-th MS, where $i \in \mathcal{S}$, receives

$$y_i = \mathbf{h}_i^T \mathbf{w}_i u_i + \sum_{j \in \mathcal{S}, j \neq i} \mathbf{h}_i^T \mathbf{w}_j u_j + n_i. \qquad (4.7)$$

The term $\sum_{j \in \mathcal{S}, j \neq i} \mathbf{h}_i^T \mathbf{w}_j u_j$ represents the detrimental ICI. In matrix notation the scheduled users receive

$$\mathbf{y} = \mathbf{H}(\mathcal{S})\mathbf{W}\mathbf{u} + \mathbf{n} \qquad (4.8)$$

where $\mathbf{y} = [y_1, \ldots, y_{|\mathcal{S}|}]$ is the received signal vector and $\mathbf{n}$ is a vector of independent complex circularly symmetric additive Gaussian noise components. The SINR γ_i of the i-th MS is

$$\gamma_i = \frac{\left|\mathbf{h}_i^T \mathbf{w}_i\right|^2 p_i}{\displaystyle\sum_{j \in \mathcal{S}, j \neq i} \left|\mathbf{h}_i^T \mathbf{w}_j\right|^2 p_j + \sigma^2}. \tag{4.9}$$

The term $\sum_{j \in \mathcal{S}, j \neq i} \left|\mathbf{h}_i^T \mathbf{w}_j\right|^2 p_j$ corresponds to the ICI power.

PAPCs are considered due to the fact that cooperating antennas are spatially distributed and they cannot share their power. It is assumed that each antenna has an average power constraint, thus $\mathbb{E}\left[|x_n|^2\right] \leq P_n$ for $n = 1, \ldots, B$. The optimal power allocation vector with respect to sum-rate maximization can be obtained by the use of an interior point method [92]. Here we consider a simpler and suboptimal equal power allocation policy. In this case $\mathbf{p} = p\mathbf{1}_{[|\mathcal{S}| \times 1]}$ and the set of constraints reduces to $\left[\mathbf{W}\mathbf{W}^H\right]_{ii} p \leq P_i$ for all $i = 1, \ldots, B$ [2]. The power allocation vector that meets these constraints is [38]

$$\left[\mathbf{W}\mathbf{W}^H\right]_{nn} p \leq P_n, \quad n = 1, \ldots, B. \tag{4.10}$$

Therefore the power allocation vector is

$$\mathbf{p} = \min_{n=1,\ldots,B} \left\{ \frac{P_n}{\left[\mathbf{W}\mathbf{W}^H\right]_{nn}} \right\} \mathbf{1}_{[|\mathcal{S}| \times 1]}. \tag{4.11}$$

As a result, the SINR of the i-th MS is

$$\gamma_i = \frac{\left|\mathbf{h}_i^T \mathbf{w}_i\right|^2}{\displaystyle\sum_{j \in \mathcal{S}, j \neq i} \left|\mathbf{h}_i^T \mathbf{w}_j\right|^2 + \sigma^2 / \min_{n=1,\ldots,B} \left\{ \dfrac{P_n}{\left[\mathbf{W}\mathbf{W}^H\right]_{nn}} \right\}}. \tag{4.12}$$

With equal power allocation and an equal power constraint per BS, $P_n = P$ for $n = 1, \ldots, B$, the expression for the power allocation vector (4.11) reduces to

$$\mathbf{p} = \frac{P}{\max\limits_{n=1,\ldots,B} \left\{ \left[\mathbf{W}\mathbf{W}^H\right]_{nn} \right\}} \mathbf{1}. \tag{4.13}$$

The entity of the network responsible for user scheduling and precoder design receives an imperfect version of the matrix $\widehat{\mathbf{H}}$ due to errors introduced by the feedback channel and due to the quantization error in case digital feedback is employed. The chosen precoding scheme is zero-forcing, where the precoding matrix inverts the imperfect channel matrix describing the received CSI. Hence the precoding matrix is

$$\mathbf{W} = \widehat{\mathbf{H}}^{H}(\mathcal{S}) \left[\widehat{\mathbf{H}}(\mathcal{S}) \widehat{\mathbf{H}}^{H}(\mathcal{S}) \right]^{-1} \mathbf{D} \qquad (4.14)$$

where $\mathbf{D}$ is a diagonal matrix that normalizes the columns of $\mathbf{W}$ to unit norm. Note that other choices of linear precoding apart from zero-forcing, e.g. the Minimum Mean Square Error (MMSE) precoder, can be considered [16]. The evaluation metric we are interested in is the average achievable rate per cell

$$\bar{C} = \frac{1}{B} \mathbb{E}_{H} \left[\sum_{i \in \mathcal{S}} \log_2 \left(1 + \gamma_i \right) \right]. \qquad (4.15)$$

4.4   Feedback Models

In this section the assumed feedback models are presented, namely analog and digital (quantized) feedback.

4.4.1   Analog Noisy Feedback

It is assumed that CSI is fed back unquantized and that a noisy version of it arrives at the target BS or BSs for both centralized and decentralized approaches. The noise process is independent on each link, and therefore in the decentralized case each BS receives a different noisy version of the CSI. Under the assumption of noisy analog feedback, each channel coefficient h_{ij} is received as follows

$$\tilde{h}_{ij} = \left(\Gamma_{ij} + w_{ij} \right) \sqrt{\mathbb{E} \left[|h_{ij}|^2 \right]} \qquad (4.16)$$

where $w \sim \mathcal{NC}\left(0, \sigma_w^2 \right)$ represents the AWGN affecting the received CSI. This inevitably degrades performance of both frameworks, as some useful information is lost by the addition of noise. This is caused as the performance of the scheduling phase is degraded due to the corrupted CSI information and also the beamforming matrix design is affected due to the same corrupted CSI. The decentralized framework can be more sensitive to scheduling degradation as inaccurate CSI might result in selection of different users by some of the cooperating BSs, depending on the scheduling algorithm employed, which will inevitably increase intra-cluster interference. However, round-robin scheduling is robust to CSI feedback errors since its scheduling decisions are not based on CSI [93].

4.4.2 Quantized Limited Feedback with Errors

In the case of quantized limited feedback, for each user i there is a quantization codebook $\mathcal{C}_i = [\mathbf{c}_1, \mathbf{c}_2, \ldots, \mathbf{c}_N]$ consisting of $N = 2^M$ vectors of unit norm, where M is the number of feedback bits. This codebook is known both by the user and by the scheduling entity. Each MS after obtaining an estimate of its channel vector $\mathbf{h}_i$ (in this chapter we assume a perfect estimate), quantizes its direction $\overline{\mathbf{h}}_i = \mathbf{h}_i / \|\mathbf{h}_i\|$ to the vector from the codebook $\mathcal{C}_i$ that best approaches it, which is the one leading to the smallest angle separation [18–21]. Therefore

$$\widetilde{\mathbf{h}}_i = \mathbf{c}_k, \quad k = \arg\max_{q=1,\ldots,N} \left| \overline{\mathbf{h}}_i^H \mathbf{c}_q \right| = \arg\max_{q=1,\ldots,N} |\cos(\angle(\mathbf{h}_i, \mathbf{c}_q))|, \quad (4.17)$$

where

$$|\cos(\angle(\mathbf{h}_i, \mathbf{c}_q))| = \frac{\left| \mathbf{h}_i^H \mathbf{c}_q \right|}{(\|\mathbf{h}_i\| \|\mathbf{c}_q\|)} \quad (4.18)$$

results from the inner product rule. The quantity determining the effectiveness of quantization is the *quantization error* defined as

$$\sin^2(\angle(\mathbf{h}_i, \mathbf{c}_k)) := 1 - \cos^2(\angle(\mathbf{h}_i, \mathbf{c}_k)). \quad (4.19)$$

The codebook should be user specific in order to avoid multiple users quantizing their channel direction to the same vector.

After quantization, MS i feeds back to the system the index k in binary form which corresponds to the quantization vector that best describes its channel direction. Therefore this piece of information is defined as Channel Direction Indicator (CDI). The more the feedback bits are, the larger the quantization codebook is, which leads to a better approximation of the MSs channel direction. Apart from CDI, the scheduling entity needs some information regarding the channel quality of each user in order to be able to make user selection decisions; this is defined as Channel Quality Indicator (CQI). In this chapter we consider the unquantized channel norm $\|\mathbf{h}_i\|$ as the fed back CQI which does not capture the inter-user interference. This is because we are interested in investigating the precoding performance and not the effectiveness of scheduling; therefore the consideration of more complex CQI metrics is unnecessary.

A *random codebook* has been considered since optimization of the codebook design is beyond the scope of our study. Thus the codebook is comprised of unit norm complex Gaussian random vectors, $\mathbf{c}_i \in\sim \mathcal{NC}\left(\mathbf{0}_B, \mathbf{I}_B\right)$ and $\|\mathbf{c}_i\| = 1$.

Consequently the concatenated quantized channel matrix of the system is $\widetilde{\mathbf{H}} = \left[\widetilde{\mathbf{h}}_1, \widetilde{\mathbf{h}}_2, \ldots, \widetilde{\mathbf{h}}_K\right]^T$. The binary indices corresponding to the CSI vectors (rows of $\widetilde{\mathbf{H}}$) are fed back to the scheduling entity through one or several radio channels (depending on the MCP framework) that introduce errors. Therefore the concatenated channel possessed by the scheduling entity is

$$\widehat{\mathbf{H}} = f\left(\widetilde{\mathbf{H}}\right) \tag{4.20}$$

where $f\left(.\right)$ is a function of the errors introduced by the feedback channel.

When quantized feedback is employed, each MS i feeds back a sequence of M bits, $\mathbf{s}_i^{Tx} = \left[s_{i1}^{Tx}, s_{i2}^{Tx}, \ldots, s_{iM}^{Tx}\right]$, $s_{ij}^{Tx} \in \{0, 1\}$. These bits are the index of the vector in the employed codebook $\mathcal{C}_i$ that best describes the CDI of a user in binary form. It is assumed that each transmitted bit s_{ij}^{Tx} is received in error with probability P_e. Therefore

$$\Pr\left\{s_{ij}^{Rx} \neq s_{ij}^{Tx}\right\} = P_e. \tag{4.21}$$

The probability of bit errors is considered to be identical and independent across different radio links. The received feedback can be protected from errors by the use of appropriate error correction techniques requiring the addition of a number of bits in the fed back sequence (see section 4.6.1).

4.5   Typical Centralized Framework for MCP

The typical conception for MCP entails that the collaborating BSs are interconnected via low latency backhaul links. These links are responsible for carrying the necessary signals that allow the cooperating BSs to act jointly; to jointly perform user scheduling and to design the transmission parameters for the scheduled users. In a linear precoding framework these parameters are the beamforming weights applied at each BS antenna of the cooperation cluster. The entity coordinating this joint action is a Control Unit accommodated in each cooperation cluster. It gathers global user CSI and centrally performs MS scheduling and signal processing operations.

In this typical centralized MCP framework, each MS is associated to a so-called *Master* or *Anchor* BS and it conceptually belongs to its corresponding cell. There are three main phases in downlink communications of FDD systems that consider incorporating MCP [6–8],

1. *Phase 1*

   - MSs estimate the CSI related to all cooperating BSs through downlink pilots. In this chapter perfect channel estimation is assumed and thus each MS i estimates the channel vector $\mathbf{h}_i$.
   - In case limited digital feedback is employed, MSs quantize the direction of their channel estimate, i.e. MS i quantizes its channel direction $\bar{\mathbf{h}}_i$ to $\tilde{\mathbf{h}}_i$.

2. *Phase 2* (Fig. 4.1)

   - *MSs feedback their CSI (CDI and CQI for digital feedback) to their Master BS* with the proper power and modulation and coding scheme (MCS) in order for the BS to be able to decode the information. All cooperating BSs gather local CSI, the CSI of the MSs belonging to their cells.
   - BSs forward the local CSI to the CU of the cluster through the low latency backhaul links. Therefore, the cluster CU collects global CSI $\widehat{\mathbf{H}} = \left[\widehat{\mathbf{h}}_1, \widehat{\mathbf{h}}_2, \ldots, \widehat{\mathbf{h}}_K\right]^T$ affected by the errors on the feedback channel.

3. *Phase 3* (Fig. 4.2)

   - The CU schedules MSs based on global CSI $\widehat{\mathbf{H}}$.
   - The CU designs the beamforming weights for each BS antenna and communicates them together with the scheduling decisions to the corresponding BSs for the transmission to take place.

This framework requires a significantly increased infrastructural cost comparing to the conventional cellular systems, as there is a demand for low latency inter-base links and a CU per cooperation cluster. Furthermore there is a need for an increased communication protocol complexity in order for these entities to interoperate properly. These facts inevitably imply changes in the current architecture of cellular systems in order for MCP to be enabled. However it is highly desirable that changes to the current structure of cellular systems are kept to a minimum when MCP capabilities are enabled.

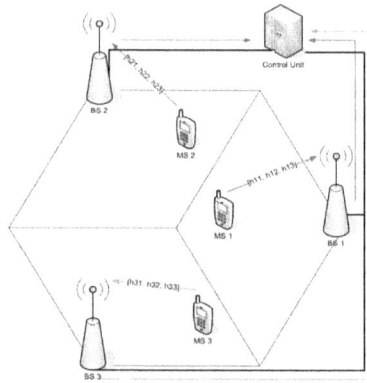

Figure 4.1: Phase 2 of the typical centralized MCP framework.

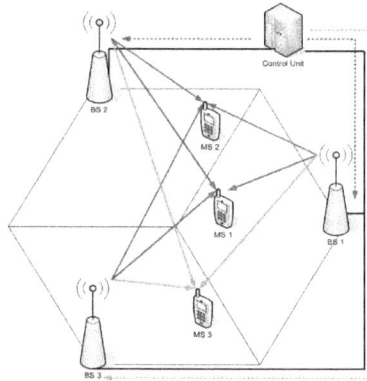

Figure 4.2: Phase 3 of the typical centralized MCP framework.

Probability of CSI error with Digital Feedback

In the centralized framework, each MS i utilizes only one radio link to transmit the bit sequence describing its quantized CSI $\mathbf{s}_i^{Tx}$, the link to the Master BS. Therefore if each bit faces an independent error probability per feedback link P_e (4.21), the probability that the i-th user's bit sequence $\mathbf{s}_i^{Rx}$ is received in error by the scheduling entity (CU) is

$$\Pr\left\{\mathbf{s}_i^{Rx} \neq \mathbf{s}_i^{Tx}\right\} = \sum_{n=1}^{M} \binom{M}{n} P_e^n \left(1 - P_e\right)^{M-n} \qquad (4.22)$$

where $\binom{K}{n}$ denotes the binomial coefficient. The bit sequence is in error if at least one of its bits is in error; thus for a specific bit error probability P_e, the more the feedback bits are, the more likely an error occurs. From this perspective it is of interest that the codebook size is kept as small as possible. However the smaller the codebook, the less accurately it can approach the actual channel state of the user, which leads to an inferior performance. Therefore a tradeoff exists between the needed codebook precision and its size with respect to feedback errors.

4.6   Proposed Decentralized Framework for MCP

In order to face the setbacks of the typical centralized framework, we propose a framework that does not require centralized scheduling and transmission design, but still can achieve the same performance. One justification for centralized processing is that the involved BSs at each cooperation cluster are assumed to lack global user CSI $\widehat{\mathbf{H}}$; they only possess local CSI which is comprised of submatrices of $\widehat{\mathbf{H}}$. MCP could be achieved in a decentralized fashion if each involved BS obtained global CSI. Therefore taking this into account, the phases of the proposed framework for the downlink are

1. *Phase 1* (identical with the centralized framework)

    - MSs estimate the CSI related to all cooperating BSs through downlink pilots. In this chapter perfect channel estimation is assumed and thus each MS i estimates the channel vector $\mathbf{h}_i$.

    - In case limited digital feedback is employed, MSs quantize the direction of their channel estimate, i.e. MS i quantizes its channel direction $\bar{\mathbf{h}}_i$ to $\widetilde{\mathbf{h}}_i$.

2. *Phase 2* (Fig. 4.3)

  - *MSs broadcast their CSI (CDI and CQI for digital feedback) to all cooperating BSs* by utilizing the radio links connecting them with the collaborating BSs. Each MS broadcasts its CSI omnidirectionally and this transmission is done with the proper power and MCS in order for all cluster BSs to be able to decode the information. All cooperating BSs gather global CSI, the CSI of the MSs of all cooperating cells.

3. *Phase 3* (Fig. 4.4)

  - The BSs schedule MSs independently based on their acquired global CSI. Cluster BSs are synchronized and employ the same scheduling algorithm. In case there are no feedback errors, BSs receive the same input parameters (global CSI $\hat{\mathbf{H}}$) and the schedulers end up selecting exactly the same MSs. If feedback links introduce errors, the fed back CSI (CDI and CQI) can be protected by the use of appropriate techniques.

  - Each BS designs the complete beamforming matrix and utilizes the antenna weights corresponding to it, i.e. BS k utilizes for transmission the k-th line of the precoding matrix $\mathbf{W}$.

Under this framework, infrastructural costs and signaling protocol complexity are minimized when MCP is enabled as neither a CU per cluster is required nor the low latency links connecting it with the cooperating BSs. Hence, the structure of MCP enabled cellular networks can remain almost the same with the structure of the conventional cellular systems. Note that under this framework, radio feedback overhead remains the same comparing to the conventional centralized framework, provided that the same resources are allocated to the terminal for feeding back its CSI by each cooperating BS.

Probability of CSI error and Discrepancy with Digital Feedback

In case errors are introduced to the fed back information, the decentralized framework can be more sensitive than the centralized one as error patterns can be different on each employed feedback link. In the centralized framework, each MS utilizes only one radio link for feeding back its CSI (CSI transmitted to the Master BS only); therefore there is only one error pattern affecting feedback information per MS in this case. In the decentralized

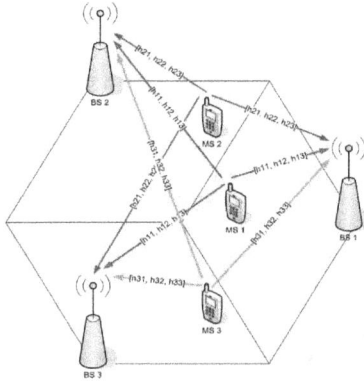

Figure 4.3: Phase 2 of the proposed decentralized MCP framework.

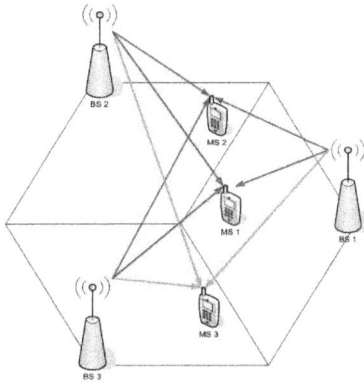

Figure 4.4: Phase 3 of the proposed decentralized MCP framework.

framework MSs feed back their CSI to all cooperating BSs, thus each BS k might acquire a different version of the global CSI $\widehat{\mathbf{H}}_k$.

Independent bit error probabilities on each feedback link increase the probability that the transmitted bit sequence of each user i, $\mathbf{s}_i^{Tx}$ is received in error in at least one of the B collaborating BSs. This probability is

$$\Pr\left\{\mathbf{s}_i^{Rx} \neq \mathbf{s}_i^{Tx}\right\} = B\left[\sum_{n=1}^{M}\binom{M}{n}P_e^n\left(1-P_e\right)^{M-n}\right]. \qquad (4.23)$$

Hence feedback errors can potentially cause a further performance degradation to the decentralized framework as compared to the centralized one (see (4.22)). Furthermore, it is interesting to investigate how close these two frameworks perform under the existence of feedback errors.

Let $\mathcal{B}$ be the set of the cooperating BSs, where $|\mathcal{B}| = B$. Independent bit error probabilities on each feedback link might result in the selection of different channel vectors by some of the B base stations for a specific MS. For example it is possible for the received feedback of BS k that $\mathbf{s}_i^{Rx,k} \neq \mathbf{s}_i^{Rx,\mathcal{J}}$, where $\mathcal{J} \subset \mathcal{B}$, $k \notin \mathcal{J}$. A set of BSs $\mathcal{J}$ might receive a different CSI index than BS k. This potential CSI discrepancy can lead to performance degradation of the decentralized framework (degradation on scheduling and precoding design). Therefore a good index for the discrepancy of the possessed CSI between the B collaborating BSs is the probability of this discrepancy to occur. Consequently this is also an index of the performance gap between the two MCP frameworks. This probability of CSI discrepancy P_d is given by the following expression

$$P_d = 1 - \sum_{m=0}^{M}\binom{M}{m}P_e^{B\cdot m}\left(1-P_e\right)^{B\cdot(M-m)}. \qquad (4.24)$$

This is due to the fact that the fed back bit sequence remains identical across the B links if the same bits in sequence are either correct or in error. In this case there is no CSI discrepancy and the probability for this event to occur P_{nd} if B base stations collaborate and M bits are transmitted is

$$P_{nd} = \sum_{m=0}^{M}\binom{M}{m}P_e^{B\cdot m}\left(1-P_e\right)^{B\cdot(M-m)}. \qquad (4.25)$$

Consequently the probability of discrepancy is $P_d = 1 - P_{nd}$ (4.24). The probability of CSI information discrepancy as a function of the number of feedback bits and the number of independent transmission links is plotted in Fig. 4.5 ($P_e = 10^{-2}$) and Fig. 4.6 ($P_e = 10^{-3}$). It can be seen that the

Probability of Bit Error = 1/100

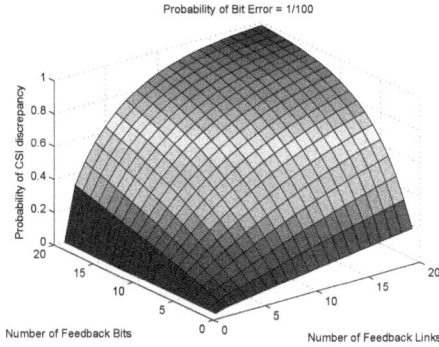

Figure 4.5: Probability of CSI discrepancy versus the number of feedback links and feedback bits ($P_e = 10^{-3}$).

probability of CSI discrepancy heavily relies on the probability of bit error P_e. Furthermore, the more the transmitted bits and the cooperating BSs are, the more likely a CSI discrepancy occurs.

In the physical layer, the bit error probability can be reduced by the use of advanced error correction techniques whereas the number of transmitted bits can be reduced by an intelligent codebook design. CSI discrepancy can also be prevented in the Medium Access Control (MAC) layer, when error detection is enabled, with the appropriate signaling techniques [94]. The impact of feedback errors is further evaluated in section 4.7.

4.6.1   Solutions for Enhancing the Robustness of Decentralized MCP

The robustness of decentralized MCP can be enhanced in practice on the physical and MAC layers by the following proposed techniques.

Malfunction Prevention Schemes

In order to robustify feedback information (CQI, ACK/NACK) against different error patterns that can be introduced by the different radio link utilised, efficient Forward Error Correction (FEC) schemes can be employed with an increased coding rate. This reduces the Block Error Rate (BLER)

Probability of Bit Error = 1/1000

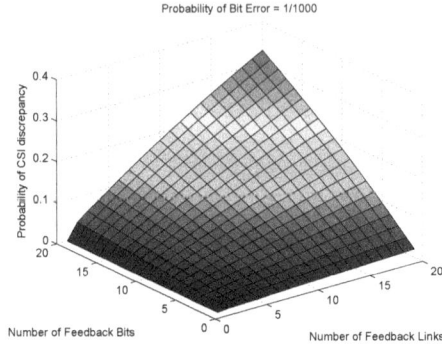

Figure 4.6: Probability of CSI discrepancy versus the number of independent links and feedback bits ($P_e = 10^{-2}$).

of the control information and therefore the occurrence of error patterns that may vary. Furthermore time diversity can also be exploited in order to augment the probability that the fed back information is correctly received by the collaborating BSs. For example feedback can be repeatedly transmitted over several Time Transmission Intervals (TTIs). Schemes that enhance feedback reliability are available in LTE specifications (e.g. ACK/NACK repetition, see [95]), which could be used in order to meet the requirements of decentralized collaborative processing. This family of schemes has the consequence of increasing the feedback overhead in the uplink.

Check Schemes

This category of schemes is responsible for ensuring that the collaborating BSs do not possess erroneous or diverging feedback information in order to avoid potential system malfunctions. If Cyclic Redundancy Check (CRC) is employed after control information encoding, the BSs can send to the users ACK/NACK signals depending on whether they have correctly received the users' HARQ messages. Thus if the MS receives ACK signals from all BSs, it then can feed back an OK signal indicating that scheduling and transmission can continue in the next time slot by taking into account this MS. On the contrary if the MS receives a NACK it feeds back a NOK signal indicating that it should not be considered for scheduling and transmission in the following time slot since there is a discrepancy regarding ACK/NACK information between the different collaborating BSs.

This discrepancy could lead to non-identical scheduling decisions and therefore to a potential system malfunction. The OK/NOK messages can be transmitted with a very high coding rate in order to maximize the probability that they are correctly received by the BSs. In case CRC is not employed by the BSs, they can just retransmit their received ACK/NACK messages to the MSs. If the MSs receive the HARQ messages they have transmitted by all collaborating BSs, they feed back an OK signal, else a NOK one. In addition to signaling overhead in both the DL (for the BS ACK/NACK transmission) and the UL (for OK/NOK transmission), this technique adds a delay of at least two TTIs to the round trip time: one TTI for the BS to transmit ACK/NACK to the UTs and another one for the UTs to transmit the OK/NOK to the BS. Processing delays may further increase this delay.

Malfunction Detection and Recovery Schemes

This group of techniques is responsible for detecting an operational malfunction and restoring stability in case the previous schemes fail to prevent the occurrence of a malfunction in the system. A potential operational anomaly may be detected either by the MSs or by the involved BSs. For instance, if the BSs receive mainly NACKS from the MSs, that might signify a problem preventing the good reception of packets targeted to specific MSs. If a system malfunction is detected, the scheduling operation needs to be restarted possibly by the use of a special signal exchanged through the backhaul links inter-connecting BSs (X2 inter-BS communication interface for LTE). In case the MS knows the cooperating BSs from which it is to receive useful signals, it can be the entity that detects an improper system function. For instance, if each collaborating BS allocates specific resources to each scheduled MS and communicates them to it, then a MS can immediately detect if a wrong number of BSs has allocated resources to it. In this case it can signal that a system error has occurred and the scheduling operation can be restarted.

4.7   Numerical Evaluation

In this section we evaluate the achievable rate performance of the proposed decentralized framework as a function of feedback errors and we compare

it with the typical centralized one. Three mutually interfering sectors of
sectorized cells (the cell radius is 1 km) have been assumed to cooperate
since this is a scenario of a particular practical interest (see figures 4.1 and
4.3). The channel coefficient between the i-th MS and the j-th sector is given
by (2.19). NLOS channels are assumed, thus multipath fading is described
by the Rayleigh distribution of (2.6). Shadowing follows the log-normal
distribution of (2.5) with standard deviation 8 dB and path-loss follows the
model of (2.4). The sector antenna power gain as a function of the horizontal
angle is given by (2.20). Transmit power is determined by the System SNR,
as detailed in section 2.7.

We assume that each MS i obtains a perfect estimate of the channel
vector associated to all cooperating BSs ($\mathbf{h}_i$). In the case of digital feedback
this estimate is quantized and then fed back omnidirectionally (CDI and
CQI feedback). In the centralized framework the feedback of each MS is
received only by its Master BS (see Fig. 4.1). In the decentralized framework
all cooperating BSs receive the CSI feedback in order for the decentralized
cooperation to take place (see Fig. 4.3).

4.7.1   Analog Feedback

Noisy feedback inevitably degrades sum-rate performance of both frameworks; both scheduling effectiveness and precoding design can be impaired
due to the corrupted CSI information. For the present evaluation round-
robin scheduling is employed as it is robust to CSI feedback errors because
scheduling decisions do not depend on CSI. If there are not any feedback
errors the performance of the two frameworks with any kind of scheduling
and transmission strategy coincides.

In Fig. 4.7 the average sum-rate performance is plotted against the noise
variance σ_w^2 of the fed back CSI for system SNR equal to 20 dB. It can be
seen that the centralized framework is slightly more robust to feedback noise.

4.7.2   Digital Feedback

Codebook Size

When quantized feedback is employed, the number of feedback bits determines precoding performance. Notably, the number of bits should increase
linearly with the SNR and the number of transmit antennas [20]. However,
the more the employed bits, the greater the probability of errors in the received CSI. This can be seen in Fig. 4.8 and Fig. 4.9, where it is plotted the
sum-rate of the system with round-robin scheduling and limited feedback as

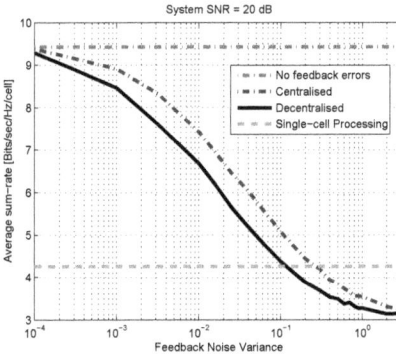

Figure 4.7: *Analog Feedback*: Sum-rate versus the feedback error noise variance for the decentralized and the centralized framework.

a function of the available bits. This is for System SNR of 10 dB and 20 dB respectively and when random codebook is employed.

   Without any feedback errors, when system SNR is set to 10 dB, 16 bits provide a good approximation of the perfect CSI, whereas for system SNR equal to 20 dB (high power regime) 20 bits are approximately needed. Notably, MCP with quantized feedback needs more quantization bits compared to the conventional limited feedback multiuser MIMO systems as it is crucial that the average capacity is sufficiently above the capacity of the single-cell processing case.

Impact of Feedback Errors

Feedback errors inevitably cause a performance degradation for both frameworks because some useful information is lost by the intervention of bit errors in the fed back quantized CSI. This is because the performance of the scheduling phase can be degraded due to the corrupted CSI information and also precoding matrix design is affected due to the same corrupted CSI. The decentralized framework can be more sensitive to scheduling degradation since imperfect CSI might result in a selection of different users by some of the cooperating BSs, depending on the scheduling algorithm employed, which will inevitably increase inter-user interference. However, round-robin scheduling is robust to CSI feedback errors since scheduling decisions are not

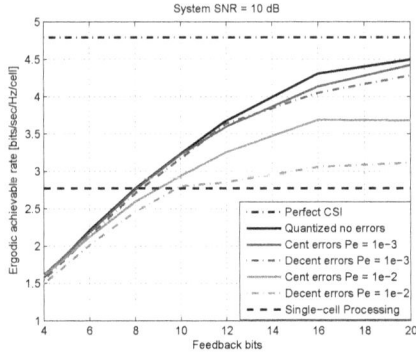

Figure 4.8: *Digital Feedback*: Sum-rate versus the number of feedback bits (System SNR = 10 dB). Blue curves: $P_e = 10^{-3}$, Red curves: $P_e = 10^{-2}$. 'Cent': centralized framework, 'Decent': decentralized framework.

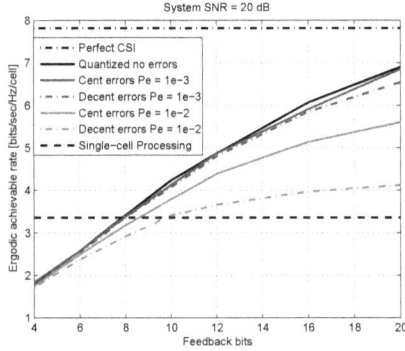

Figure 4.9: *Digital Feedback*: Sum-rate versus the number of feedback bits (System SNR = 20 dB). Blue curves: $P_e = 10^{-3}$, Red curves: $P_e = 10^{-2}$. 'Cent': centralized framework, 'Decent': decentralized framework.

made based on CSI. This scheduling algorithm is selected for the present evaluation which focuses on the impact of feedback errors on the design of precoding matrices. Note that with the absence of feedback errors the performance of the two frameworks with any kind of scheduling and transmission strategy coincides.

In Fig. 4.10 and Fig. 4.11 the average capacity is plotted against the probability of bit errors P_e when MSs are scheduled in a round-robin fashion for system SNR of 10 and 20 dB respectively. When the system SNR is 10 dB, 16 bits are chosen for feedback whereas at 20 dB, 20 feedback bits are considered. This increase in bits is justified by the higher power regime of operation which demands larger codebooks for guaranteeing low inter-user interference [20]. It can be seen that feedback errors have a much greater impact in the case of 20 bits which is justified by (4.22) and (4.23). Furthermore in the case of 20 bits the performance gap between the decentralized and centralized frameworks is increased, an increase justified by the greater probability of CSI discrepancy (4.24). Therefore a more intelligent codebook design can lead to a reduced number of feedback bits for the same performance targets and also provide better robustness against feedback errors.

Generally, the centralized framework is a little more robust to feedback errors than the decentralized one, although for a bit error probability less than 10^{-3} the difference is negligible. It should also be noted that the presented results reflect the case where no error detection/correction schemes have been employed. These schemes can significantly reduce the bit error probability P_e, thus they have the potential to eliminate the impact of feedback errors.

4.8   Conclusions

MCP promises significantly improved spectral efficiency and fairness for future cellular systems. However, this comes at the cost of increased infrastructural complexity and centralized processing related to the existing conception for MCP. According to this, cooperating BSs need to be connected to a CU which plays the role of the cluster head. It gathers local CSI from the BSs, performs user scheduling and designs the transmission parameters.

In this chapter a new framework has been proposed that allows MCP in the downlink to take place in a decentralized fashion; neither a CU nor the low latency backhaul links are needed. Each BS receives CSI feedback from all the users of the cluster (global CSI) and designs transmission independently. The performance of the proposed framework has been evaluated

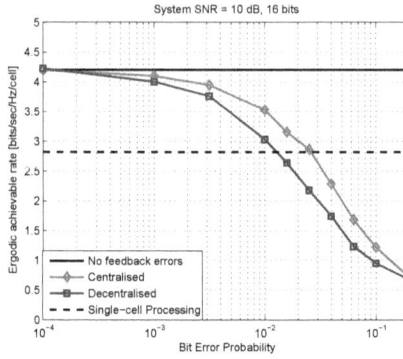

Figure 4.10: *Digital Feedback*: Sum-rate versus bit error probability for both frameworks (System SNR = 10 dB, 16 bits employed).

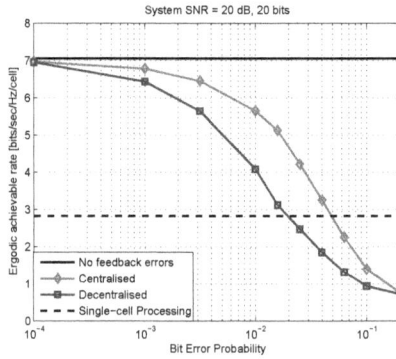

Figure 4.11: *Digital Feedback*: Sum-rate versus bit error probability for both frameworks (System SNR = 10 dB, 20 bits employed).

under analog and quantized digital feedback together with linear precoding and while feedback errors are introduced by the channel. It has been shown that the proposed framework shows little degradation on the achievable sum-rate compared to the centralized alternative, which can be eliminated with a more intelligent codebook design and the addition of feedback protection schemes (error detection/correction). The decentralized framework allows MCP to be implemented with very few changes upon the current network architecture.

Chapter 5

Limiting Feedback and Backhaul Overheads of MCP

In the previous chapters, the challenges of Multicell-MIMO enabled systems related with the BS cluster size and the infrastructural overheads have been addressed. In addition to these overheads, MCP systems heavily suffer from feedback and backhaul overheads, and these aspects are investigated in this chapter.

The throughput of a downlink multiuser MIMO channel heavily relies on the quality of the Channel State Information (CSI) available at the transmitter [61, 62, 96]. Therefore in FDD systems employing multiuser MIMO techniques (single-cell processing), MSs need to estimate the downlink CSI and feed it back to the network infrastructure in order for the downlink precoder to be computed. We refer to this as *feedback overhead*. This overhead is more dominant in MCP systems as users need to estimate and feed back at least as many channel coefficients as the number of cooperating antennas. Furthermore, in these networks each BS potentially transmits to an increased number of users at a time (space divison multiple access - SDMA) and thus needs to buffer an increased number of data streams. We will hereby refer to this as *backhaul overhead*. Consequently MCP enabled systems require backhaul links of higher capacity which implies an elevated deployment cost. Therefore reducing backhaul load by routing packets only to BSs that really need it, is desirable because it can lead to deployment cost reduction. In addition it can make the use of MCP possible in scenarios where it was not initially considered feasible due to high bit rate constraints for cell-to-cell signaling.

Techniques for reducing CSI feedback load in multiuser MIMO systems, where MCP is not enabled, have been researched extensively [66–69, 97]. In the MCP case, feedback load is much greater since MSs need to estimate the CSI related to all BSs that cooperate and feed it back to the system infrastructure. Techniques for reducing feedback overhead for MCP [10, 12] and backhaul overhead [45, 47] have been already investigated but without attempting to jointly minimize the two.

In this chapter a technique for overhead reduction in the multicell context based on the use of user selective feedback is proposed targeted at both the mitigation of feedback and backhaul overheads. *Selective feedback* has been previously introduced in the context of single-cell processing and targeted at multiuser diversity [66, 67]. In the MCP context, selective feedback relies on users estimating their downlink channel seen from surrounding BSs and deciding, on the basis of the comparison with a pre-determined threshold, whether they should engage in MCP or not. More specifically, we propose an algorithm according to which each MS feeds back to the system infrastructure the channel coefficients whose average SNR is above an absolute threshold, in order to keep feedback load at prescribed target levels. The multicell setting impacts the channel statistics as channels to different BSs undergo different pathloss and large-scale fading (shadowing). The feedback load as a function of a chosen SNR threshold is studied analytically and this analysis differs from [66, 67] in that the fed back channel coefficients are associated with more than one BSs, hence they have different power profiles. The other main point made in this chapter is how the feedback reduction can be combined with the reduction of the inter-base backhaul overhead.

5.1   System Model

A network consisting of B single antenna BSs and K single antenna active MSs overall is considered. The assumption of single antenna BSs is mostly for expository reasons and does not preclude applying the same concepts to multiple antenna BSs. In the present paper we focus in the downlink but similar ideas could be applied in the uplink. Furthermore, flat fading and spatio-temporally uncorrelated channels are assumed. The k-th user of the system receives

$$y_k = \mathbf{h}_k^T \mathbf{x} + n_k \tag{5.1}$$

where $\mathbf{h}_k = [h_{k1}, h_{k2}, \ldots, h_{kB}]^T$ is the channel vector corresponding to the k-th user, $\mathbf{x} \in \mathbb{C}^{B \times 1}$ is the vector containing the transmit signals sent by all the antennas of the network and $n_i \sim \mathcal{NC}\left(0, \sigma^2\right)$ represents the independent complex circularly symmetric additive Gaussian noise. The complete channel matrix of the system is

$$\mathbf{H} = [\mathbf{h}_1, \mathbf{h}_2, \ldots, \mathbf{h}_K]^T. \tag{5.2}$$

It is assumed that the system operates in FDD mode and that each MS k obtains a perfect estimate of its CSI consisting of the vector of channel coefficients $\mathbf{h}_k$. In order for the feedback load to be reduced, not all the coefficients of the vector $\mathbf{h}_k$ are fed back to the network infrastructure. The coefficients whose channel gain is below a specified threshold are replaced with zeros, as it is detailed in section 5.2, and this new vector $\hat{\mathbf{h}}_k$ is fed back to the network infrastructure through delayless and errorless feedback links.

Let $\mathbf{v}_k \in \{0, 1\}^{[B \times 1]}$ be the vector indicating which coefficients are fed back and which are not by the k-th user (positions of 1s and 0s respectively), e.g. $\mathbf{v}_k = [1, 0, \ldots, 1, 0, 1]^T$ [1]. It is assumed that each MS k always feeds back the strongest coefficient of the vector $\mathbf{h}_k$, thus the vector $\mathbf{v}_k$ contains at least one 1. In the limiting case where MS k feeds back its entire CSI vector $\mathbf{v}_k = \mathbf{1}_{[B \times 1]}$. Let the feedback index matrix $\mathbf{V}$ be the concatenation of all feedback index information across all users

$$\mathbf{V} = [\mathbf{v}_1, \mathbf{v}_2, \ldots, \mathbf{v}_K]^T, \quad \text{e.g.} \quad \mathbf{V} = \begin{bmatrix} 1 & 1 & \ldots & 1 & 0 \\ 0 & 1 & \ldots & 0 & 1 \\ \vdots & \vdots & \ddots & \vdots & \vdots \\ 0 & 1 & \ldots & 1 & 1 \end{bmatrix}. \tag{5.3}$$

Hence the acquired imperfect CSI matrix by the network infrastructure is of the form

$$\hat{\mathbf{H}} = \mathbf{H} \odot \mathbf{V} \tag{5.4}$$

where $\hat{\mathbf{H}} = \left[\hat{\mathbf{h}}_1, \hat{\mathbf{h}}_2, \ldots, \hat{\mathbf{h}}_K\right]^T$ and $\hat{\mathbf{h}}_k = \mathbf{h}_k \odot \mathbf{v}_k$. If a set of MSs $\mathcal{S}$ is scheduled for transmission in a specific time slot, their limited CSI matrix is $\hat{\mathbf{H}}(\mathcal{S}) = \mathbf{H}(\mathcal{S}) \odot \mathbf{V}(\mathcal{S})$.

---

[1] A zero at the j-th position means that the channel coefficient to BS j is not fed back to the system infrastructure.

In this chapter the linear precoding model introduced in section 4.3 has been employed. It is assumed that the number of scheduled users always equals the number of BSs, $|\mathcal{S}| = B$ The considered performance metric is average sum-rate per cell given by

$$\bar{C} = \frac{1}{B}\mathbb{E}_H \left[\sum_{k \in \mathcal{S}} \log_2 \left(1 + \gamma_k \right) \right]. \tag{5.5}$$

where the expectation is taken over all channel realizations and MS locations.

5.1.1   Feedback Overhead

Let $N\left(t\right)$ be the total number of channel coefficients fed back by the users in time t. Since each MS feeds back at least the strongest coefficient describing its channel state (matrix $\mathbf{V}$ contains at least K ones) $N\left(t\right) \in [K, BK]$. The feedback load reduction may be expressed by the average number of channel coefficients fed back per MS

$$\bar{L} = \frac{1}{K}\mathbb{E}_t \left[N\left(t\right) \right] \tag{5.6}$$

where $\bar{L} \in [1, B]$ as each user feeds back at least one channel coefficient and up to B.

5.1.2   Backhaul Overhead

Under a linear precoding framework and with single antenna BSs and MSs, the m-th row vector of the precoding matrix $\mathbf{W}$ corresponds to the weights applied to transmit symbols by the m-th BS, where $m = 1, \ldots, B$. The i-th weight of this vector (w_{mi}), where $i = 1, \ldots, B$ $(|\mathcal{S}| = B)$, is applied to the symbol intended for MS i. If this element is 0, BS m allocates zero power to MS i, hence it does not need to buffer the symbol s_i intended for MS i. This observation can lead to the reduction of backhaul overhead.

Let $Z\left(t\right)$ be the number of streams transmitted per BS in time slot t, $Z\left(t\right) \in [1, B]$. This corresponds to the number of non-zero elements per line vector of the precoding matrix $\mathbf{W}$. Backhaul overhead can be measured by the average number of data streams that each BS transmits per time slot per resource block

$$\bar{S} = \mathbb{E}_t \left[Z\left(t\right) \right]. \tag{5.7}$$

5.2   Feedback Overhead Reduction

In this section the performance of the feedback overhead reduction scheme is estimated analytically. In each time slot each MS k obtains a perfect estimate of the vector of channel coefficients to all BSs $\mathbf{h}_k$. The average SNR of a channel coefficient h_{kn} is defined as

$$\bar{\gamma}_{kn} = \mathbb{E}\left[\frac{|h_{kn}|^2 p_k}{\sigma^2}\right] \qquad (5.8)$$

where expectation is taken over the statistics of fast fading only and p_k is the transmit power allocated to user k. Algorithm 3 is formulated where each MS k feeds back to the system infrastructure the vector $\hat{\mathbf{h}}_k$ comprised only by the channel coefficients whose corresponding average SNR exceeds a threshold γ_t. Note that each MS feeds back at least its strongest channel coefficient upon which no threshold is applied. This permits the scheduler to make decisions upon the data streams to be transmitted by each BS, hence it also permits data routing to take place on a realistic time scale, longer than that of multipath fading.

Algorithm 3 Radio Feedback Load Reduction

Require: Define SNR threshold γ_t
1: **for** all MSs $k = 1, \ldots, K$ **do**
2:    Initialize $\mathbf{v}_k = \mathbf{0}_{[B \times 1]}$
3:    **Find** BS $n = \arg \max\limits_{n=1,\ldots,B} |h_{kn}|^2$ (BS providing the strongest channel gain to MS k)
4:    Set $v_{kn} = 1$ (coefficient h_{kn} is fed back to the system infrastructure)
5:    **for** all BSs $j = 1, \ldots, B$, $j \neq n$ **do**
6:       **if** $\bar{\gamma}_{kj} = \mathbb{E}\left[\frac{|h_{kj}|^2 p_k}{\sigma^2}\right] \geq \gamma_t$ **then**
7:          Set $v_{kj} = 1$ (h_{kj} is fed back to the system infrastructure)
8:       **else**
9:          Set $v_{kj} = 0$ (h_{kj} is not fed back to the system infrastructure)
10:       **end if**
11:    **end for**
12: **end for**
13: The acquired CSI for user k by the system infrastructure is $\hat{\mathbf{h}}_k = \mathbf{h}_k \odot \mathbf{v}_k$

5.2.1   Feedback Overhead Reduction Analysis

Let the channel coefficient between the k-th MS and the j-th BS be modeled as in (2.19). The transmission power is determined by the System SNR which is defined as the average SNR received at the edge of the cell.

Here we study analytically the average feedback load defined in (5.6). Let $P_{\bar{\gamma}}(\gamma)$ denote the CDF of the average SNR $\bar{\gamma}$ of each channel coefficient. If $N(t) = n$, it is implied that the average SNR of n out of BK channel coefficients is above the defined SNR threshold γ_t and that the average SNR of the rest $BK - n$ coefficients is below this threshold. In a multicell scenario each channel coefficient experiences a different average SNR due to the difference in pathloss and shadowing related to different BSs; therefore the global distribution followed by the channel coefficients is hard to derive in exact form. Instead we suggest the following approach; the average SNR distribution of $\bar{\gamma}_{kn}$ can be empirically well approximated by a log-normal distribution. The CDF of the log-normal distribution is

$$P_{\bar{\gamma}}(x) = \frac{1}{2}\left[1 + \mathrm{erf}\left(\frac{\ln x - \mu_{\bar{\gamma}}}{\sqrt{2}\sigma_{\bar{\gamma}}}\right)\right] \qquad (5.9)$$

where $\mu_{\bar{\gamma}}$ and $\sigma_{\bar{\gamma}}$ are the logarithmic mean and standard deviation respectively that can be obtained by using standard fitting techniques [98]. $\mathrm{erf}(x)$ is the error function defined as

$$\mathrm{erf}(x) = \frac{2}{\sqrt{\pi}}\int_0^x e^{-t^2}dt. \qquad (5.10)$$

$N(t)$ follows a binomial distribution and its probability mass function (the probability that n out of the BK channel coefficients are fed back) is

$$\Pr\{N(t) = n\} = \binom{BK}{n}(1 - P_{\bar{\gamma}}(\gamma_t))^n (P_{\bar{\gamma}}(\gamma_t))^{BK-n}. \qquad (5.11)$$

Therefore the expected number of $N(t)$ is

$$\begin{aligned}\mu_N &= \mathbb{E}\{N(t)\} = \sum_{n=0}^{BK} n\Pr\{N(t) = n\} \\ &= BK[1 - P_{\bar{\gamma}}(\gamma_t)].\end{aligned} \qquad (5.12)$$

Hence, the average number of fed back coefficients per user is

$$\bar{L} = \frac{1}{K}\mu_N = B\left[1 - P_{\bar{\gamma}}(\gamma_t)\right]. \tag{5.13}$$

By plugging (5.9) to (5.13) we obtain the following expression for the average number of fed back channel coefficients per MS as a function of the SNR threshold γ_t

$$\bar{L} = B\left[\frac{1}{2} - \frac{1}{2}\mathrm{erf}\left(\frac{\ln\gamma_t - \mu_{\bar{\gamma}}}{\sqrt{2}\sigma_{\bar{\gamma}}}\right)\right]. \tag{5.14}$$

Since the average number of fed back coefficients per MS is expressed as a function of the SNR threshold γ_t, we can design the power threshold in order to achieve specific average feedback load targets

$$\ln(\gamma_t) = \mu_{\bar{\gamma}} + \sqrt{2}\sigma_{\bar{\gamma}}\mathrm{erf}^{-1}\left(1 - \frac{2\bar{L}}{B}\right). \tag{5.15}$$

As an indicative example, the followed global PDF of the channel coefficients' average SNR, for the practical case of 3 mutually interfering sectors as shown in Fig. 5.1, is well approximated by the log-normal distribution. This approximation can be seen in Fig. 5.2 (the evaluation parameters of section 5.4 are used). In Fig. 5.3 the average feedback load $\bar{L}$ is plotted against the threshold γ_t for different values of system SNR. It can be seen that the greater the system SNR is, and thus the transmit power, the higher the feedback load becomes for a specific threshold value. Theoretical results are shown by the dashed lines while numerical results are shown by the solid ones. It can be seen that numerical results can be very well approximated by the model presented above.

5.2.2   Feedback Variation Prediction

Another important quantity is the variation of the total number of fed back channel coefficients as it permits the feedback channel to be provisioned for appropriately. The variation of $N(t)$ can be defined as

$$V_N = \frac{\mathrm{var}\left[N(t)\right]}{\mathbb{E}\left[N(t)\right]^2} = \frac{\mathbb{E}\left[(N(t) - \mu_N)^2\right]}{\mathbb{E}\left[N(t)\right]^2} \tag{5.16}$$

where

$$\mathrm{var}\left[N(t)\right] = BK\left(1 - P_{\bar{\gamma}}(\gamma_t)\right)P_{\bar{\gamma}}(\gamma_t),$$
$$\mathbb{E}\left[N(t)\right]^2 = \left[BK\left(1 - P_{\bar{\gamma}}(\gamma_t)\right)\right]^2. \tag{5.17}$$

This results to the following expression for the variation of radio feedback

$$V_N = \frac{P_{\bar\gamma}(\gamma_t)}{KB(1 - P_{\bar\gamma}(\gamma_t))} = \frac{1 + \mathrm{erf}\left(\frac{\ln\gamma_t - \mu_\gamma}{\sqrt{2}\sigma_\gamma}\right)}{KB\left(1 - \mathrm{erf}\left(\frac{\ln\gamma_t - \mu_\gamma}{\sqrt{2}\sigma_\gamma}\right)\right)}. \qquad (5.18)$$

5.3   Backhaul Overhead Reduction

According to the proposed feedback load reduction framework, MSs estimate and feed back to the system infrastructure a limited number of channel coefficients in each time slot depending on a SNR threshold. Which channel coefficients are fed back by each MS depends on their associated average SNR that is a function of the MS location due to pathloss and large-scale fading.

The chosen precoding scheme is Zero-Forcing, where the precoding matrix inverts the imperfect channel matrix describing the received CSI. Hence the precoding matrix is

$$\begin{aligned} \mathbf{W} &= [\mathbf{H}(\mathcal{S}) \odot \mathbf{V}(\mathcal{S})]^{-1}\mathbf{D} \\ &= \hat{\mathbf{H}}(\mathcal{S})^{-1}\mathbf{D} \end{aligned}$$

where $\mathbf{D}$ is a diagonal matrix that normalizes the columns of $\mathbf{W}$ to unit norm. As we are interested in guaranteeing fairness, scheduling does not aim to provide multiuser diversity gains. Our focus is on multiplexing gains which is a sensible design in the large SNR regime. The over-the-air feedback load reduction can be exploited for achieving backhaul load reduction (BLR). The following possibilities exist regarding scheduling.

5.3.1   Round-robin Scheduling without BLR

Let $\mathcal{S}$ be the set of users selected in a round-robin manner in a specific time slot. Generally there might be zero elements in random positions of $\hat{\mathbf{H}}(\mathcal{S})$ apart from its main diagonal; $\hat{\mathbf{H}}(\mathcal{S})$ in the low feedback load regime is in principle a sparse matrix. In the limiting case, where all MSs feed back just their strongest coefficients, $\hat{\mathbf{H}}(\mathcal{S})$ is a diagonal matrix and this corresponds to single-cell processing (absence of MCP). If the zero elements of $\hat{\mathbf{H}}(\mathcal{S})$ are in random positions there might be some zero elements in its inverse, $\mathbf{W}$, although their number will be smaller than the one of $\hat{\mathbf{H}}(\mathcal{S})$ and their position cannot be predicted in a straighforward manner [99] (impact on backhaul overhead). Therefore the main disadvantage of this pure round-robin approach with Zero-Forcing is that all the B collaborating BSs need to buffer as many data streams per slot as the number of serving MSs $|\mathcal{S}|$ (which is assumed equal to B). Therefore in this case the maximum average backhaul load per BS is

$$\bar{S} = B. \tag{5.19}$$

5.3.2   Scheduling for BLR

Pure round-robin scheduling is unable to mitigate backhaul load. A scheduling algorithm that can translate feedback load reduction into backhaul overhead reduction is formed by selecting a suitable set of users $\mathcal{S}$. A suitable set of users $\mathcal{S}$ is obtained if the channel matrix $\mathbf{H}(\mathcal{S})$ to be inverted has a block-diagonal structure or a structure equivalent to block-diagonal. A matrix $\mathbf{H}(\mathcal{S})$ is equivalent to a block-diagonal one if

$$\boldsymbol{\pi}^r \mathbf{H}(\mathcal{S}) \boldsymbol{\pi}^c = \mathrm{diag}(\mathbf{H}_{11}, \mathbf{H}_{22}, \dots, \mathbf{H}_{NN}) \tag{5.20}$$

where $\mathbf{H}_{nn}$ are submatrices of $\mathbf{H}$ and N is the number of blocks of possibly different sizes. $\boldsymbol{\pi}^r = \pi_1^r \pi_2^r \dots \pi_N^r$ and $\boldsymbol{\pi}^c = \pi_1^c \pi_2^c \dots \pi_M^c$ represent row and column permutations respectively. An important property of block-diagonal matrices is that their inverse of is also block-diagonal and containing the same number of zeros. This can be exploited for limiting the backhaul load of a system employing reduced feedback overhead.

In order to explain how the block-diagonal structure of $\mathbf{H}(\mathcal{S})$ can be achieved, we introduce the concept of BS subgroups. Let $\mathcal{D} = \{\mathbf{d}_1, \dots, \mathbf{d}_{|\mathcal{D}|}\}$ be the set of all potential BS subgroups, each one intended to serve users which have fed back the coefficients corresponding to the same BSs, where $|\mathcal{D}| = \sum_{k=1}^{B} \binom{B}{k}$. Let $\mathbf{d}_i \in \{0,1\}^{[B \times 1]}$ be a vector in which the position of ones indicate which BSs participate in subgroup i. Thus each subgroup described by $\mathbf{d}_i$ consists of a set of BSs which jointly transmit to a set of users (set of users aiming at receiving useful signals from these BSs). Let $z_i \in [1, B]$ contain the number of ones in each vector $\mathbf{d}_i$, $i = 1, \dots, |\mathcal{D}|$ which represents the number of BSs of subgroup i. We assume that the BS subgroups transmitting in each time slot are disjoint (each transmitting BS belongs only to one cluster).

As detailed in Algorithm 4, in each time slot a MS set related to each BS subgroup (they have fed back the channel coefficients of the subgroup's

BSs) is selected in a round-robin manner. Then a group of $N \leq B$ disjoint BS subgroups is formed, also in round-robin fashion, and its associated users are served in this time slot. In this way the channel matrix $\hat{\mathbf{H}}(\mathcal{S})$ either is or it can be converted to a block diagonal matrix with appropriate row and/or column permutations. Therefore the inverse of $\mathbf{H}(\mathcal{S})$ is also a block-diagonal matrix and in this way every BS needs to transmit a limited number of data streams in each time slot (contrary to the previous section). The average number of transmitted data streams per time slot is directly proportional to $\bar{L}$

$$\bar{S} = \bar{L} \quad \text{where} \quad \bar{S} \in [1, B]. \tag{5.21}$$

Algorithm 4 Scheduling for backhaul load reduction

1: Let $\mathcal{K}_i$ be the set of users of the i-th BS subgroup (users that have fed back the channel coefficients of the BSs of this subgroup), $i = 1, \ldots, |\mathcal{D}|$
2: **for** all MSs $k = 1, \ldots, K$ **do**
3:     **for** all BS subgroups $\mathbf{d}_i$, $i = 1, \ldots, |\mathcal{D}|$ **do**
4:         **if** $\mathbf{v}_k = \mathbf{d}_i$ **then**
5:             MS $k \in \mathcal{K}_i$ (MS k is grouped in subgroup i)
6:         **end if**
7:     **end for**
8: **end for**
9: **for** all BS subgroups $i = 1, \ldots, |\mathcal{D}|$ **do**
10:     Select a set of users $\mathbf{U}_i \in \mathcal{K}_i$, where $|\mathbf{U}_i| = z_i$ (the cardinality of the user set should equal the one of the BS subgroup), in a round-robin manner.
11: **end for**
12: Select a group of N disjoint BS subgroups also in a round-robin manner (the MSs $\mathcal{S}$ to be served are related to these subgroups).
13: The matrix to be inverted either is block-diagonal or it is converted to a block-diagonal matrix with the application of row and/or column permutations, $\boldsymbol{\pi}^r = \pi_1^r \pi_2^r \ldots \pi_N^r$ and/or $\boldsymbol{\pi}^c = \pi_1^c \pi_2^c \ldots \pi_M^c$ respectively, such that $\boldsymbol{\pi}^r \mathbf{H}(\mathcal{S}) \boldsymbol{\pi}^c = \text{diag}(\mathbf{H}_{11}, \mathbf{H}_{22}, \ldots, \mathbf{H}_{NN})$.
14: The precoder is: $\mathbf{W} = \left[\hat{\mathbf{H}}(\mathcal{S}) \right]^{-1} \mathbf{D} = \text{diag}\left(\mathbf{H}_{11}^{-1}, \mathbf{H}_{22}^{-1}, \ldots, \mathbf{H}_{NN}^{-1} \right) \mathbf{D}$.

5.3.3   BLR via Partial Inversion

Before we presented a scheduler that can achieve backhaul load reduction by selecting users who have specific feedback patterns. Here we provide an alternative approach that translates selective feedback gains to gains on the backhaul without any specific constraints on the scheduler (here we consider round-robin scheduling).

Each MS is assigned to the BS providing the strongest channel. Let $\mathcal{S}$ be the set of users selected in a specific time slot. Each BS i considers the indices $\mathcal{M}_i \subseteq \mathcal{S}$ of the users that have fed back the channel coefficient related to it. For transmission it utilizes the vector $\mathbf{w}_i$ which is taken as the i-th row extracted from the matrix $\mathbf{W}_i = \hat{\mathbf{H}}^H (\mathcal{M}_i) \left[\hat{\mathbf{H}} (\mathcal{M}_i) \hat{\mathbf{H}}^H (\mathcal{M}_i) \right]^{-1} \mathbf{D}_i$ where $\hat{\mathbf{H}} (\mathcal{M}_i)$ is a submatrix of $\hat{\mathbf{H}} (\mathcal{S})$. In this way the backhaul load is reduced as the inversion of submatrices of $\hat{\mathbf{H}} (\mathcal{S})$ implies that each BS needs to transmit less user symbols than B, thus less user streams need to be exchanged between the BSs of the network. The average backhaul load is

$$\bar{S} = \bar{L} \quad \text{where} \quad \bar{S} \in [1, B]. \tag{5.22}$$

This is identical with the expression (5.21) of the previous section, although as it can be seen in the numerical results, a precoding matrix with block-diagonal structure (disjoint BS clusters) provides superior capacity comparing to this approach.

5.4   Numerical Results

A scenario of a particular practical interest is mitigating interference of mutually interfering sectors as shown in Fig. 5.1. In this scenario the 3 mutually interfering sectors can cooperate under the reduced feedback load regime and jointly serve MSs in their area of coverage. The channel coefficient between the k-th MS and the j-th sector is modeled as in (2.19) and the shadowing is assumed to follow a log-normal distribution $\gamma_{dB} \sim \mathcal{N} (0\,dB, 8\,dB)$. The sector antenna power gain as a function of the horizontal angle ϕ in degrees is given by (2.20) and the pathloss by (2.4). The transmission power is determined by the concept of system SNR (section 2.7).

5.4.1   Feedback Load Performance

The distribution of the average SNR of the channel coefficients for the case of Fig. 5.1 has been approximated by a log-normal distribution. The average

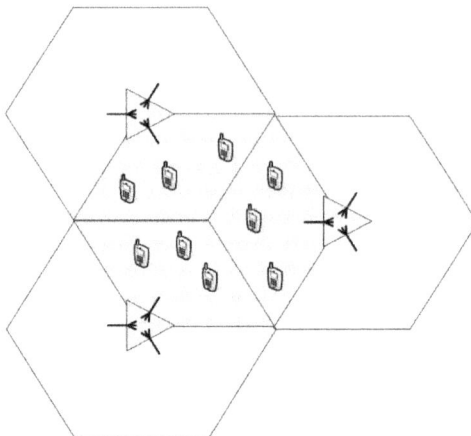

Figure 5.1: A scenario of particular practical importance: 3 mutually interfering sectors.

channel coefficient SNR distributions for System SNRs of 0 dB, 10 dB, 20 dB and 30 dB are approximated by log-normal distributions with means $\mu_{0dB} = -0.8$, $\mu_{10dB} = 1.4$, $\mu_{20dB} = 3.8$ and $\mu_{30dB} = 6.1$ respectively and standard deviation $\sigma = 2.1$ for all System SNRs (Fig. 5.2). In Fig. 5.3 the average number of fed back coefficients per MS $\bar{L}$ is plotted against the SNR threshold γ_t for various values of system SNR (5.6). It can be seen that the theoretical approximation (dashed lines) matches well the numerical results (solid lines).

5.4.2   Sum-rate Performance

In Fig. 5.4 it is plotted the average sum-rate per cell as a function of the radio feedback $\bar{L}$ (5.6) for the abovementioned scheduling approaches and for System SNR of 20 dB (interference limited regime). The solid curve corresponds to the capacity of round-robin scheduling without BLR of section 5.3.1, the dotted curve corresponds to the scheduling for BLR (the precoding matrix is block-diagonal, section 5.3.2) and the dashed curve corresponds to the case of BLR via partial inversion (section 5.3.3). It can be seen that the round-robin scheduling without BLR provides better capacity than the other

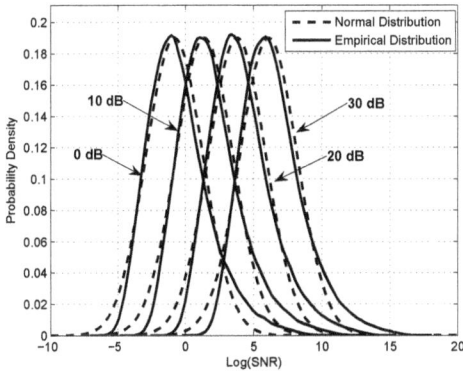

Figure 5.2: The empirical PDFs of the logarithmic average SNR of the channel coefficients for System SNR of 0 dB, 10 dB, 20 dB and 30 dB

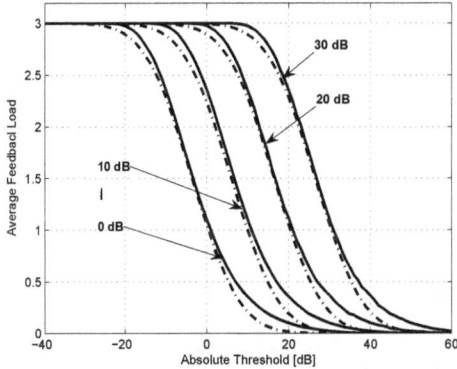

Figure 5.3: Average feedback per MS versus threshold for different values of system SNR (0 dB, 10 dB, 20 dB, 30 dB). Theoretical results are shown by the dashed lines and numerical ones by the solid ones.

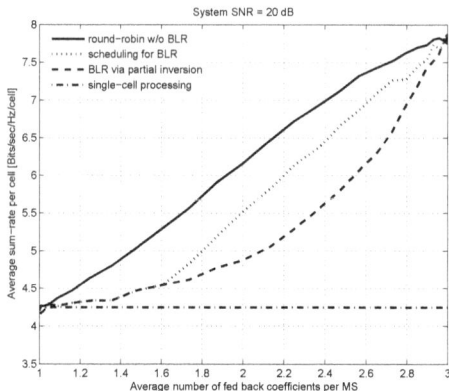

Figure 5.4: Sum-rate against feedback for the framework providing fairness.

approaches as this scheme leads to more degrees of freedom for the transmission; the zeros in the $\mathbf{H}(\mathcal{S})$ matrix to be inverted can be in any position apart from the main diagonal. However under this scheme, the feedback load savings do not translate into savings on the backhaul exchanges. The scheduling framework for BLR performs worse than round-robin scheduling but better than BLR via partial inversion approach. It should be noted that the reduction on the backhaul charge achieved by the approaches of sections 5.3.2 and 5.3.3 is directly proportional to the radio feedback load and this is the main advantage of these methods. In Fig. 5.5 it is plotted the empirical CDF of the instantaneous rate of each user for all approaches when the threshold is 12 dB and leads to the same conclusions as Fig. 5.4.

5.5   Conclusion

Multicell cooperative processing although very promising for future cellular systems, comes at the cost of increased feedback and backhaul overhead. In the downlink of FDD systems, MSs need to estimate and feed back several channel coefficients and BSs need to exchange user data. This creates the need for reducing these overheads in order for MCP to be brought into practice. In this paper a feedback load reduction technique has been

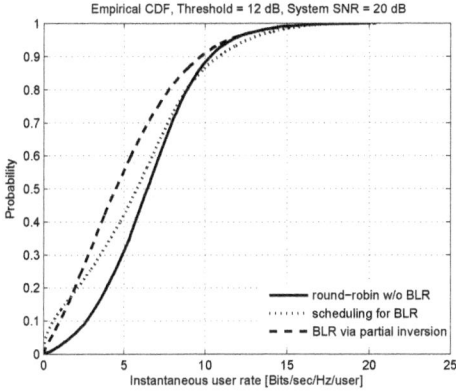

Figure 5.5: CDF of the instantaneous user rate.

proposed based on SNR thresholds and its effects have been analytically approximated. Furthermore it has been shown that reduction on the over-the-air feedback load can be efficiently exploited for reducing backhaul load through either scheduling or precoding. The proposed techniques achieve a good tradeoff between performance and complexity that can facilitate the introduction of MCP capabilities in conventional cellular systems.

Chapter 6

Utilizing Dynamic Relays in Cellular Systems

6.1 Introduction

While the previous chapters investigate the benefits and the challenges of MCP, this chapter is dedicated to cooperative relaying in cellular systems. It is well known that cooperative relaying can exploit the spatial diversity inherent in wireless systems, offering increased capacity, fairness and coverage, under several resource constraints [48, 50]. However, the utilization of relay stations (RSs) in cellular environments still remains a challenging task due to power limitations and high implementation complexity [71, 80]. Utilization of dynamic RSs is acknowledged to be more cost efficient as relay nodes are not elements of the network infrastructure but user terminals which can relay signals intended for other users [78, 79]. Their topology changes in time as users move and on one hand this hardens the process of relay selection. On the other hand, users' mobility provides a significant advantage as multi-user diversity can be exploited for increasing relaying efficiency and system performance [76].

We focus on leveraging dynamic RSs under *dual-hop* transmission. RS selection and MS scheduling are performed in a centralized fashion. In this setting the BS makes the decisions on MS and RS scheduling. The optimal strategy for maximizing performance is that all MSs of the cell are considered as potential relay nodes or destinations [76, 100]. This inevitably entails high feedback overhead and scheduling complexity as the BS needs to know all the BS-MS and MS-MS channel coefficients and perform exhaustive search in order to identify the best MS-RS pair with respect to the considered metric. Therefore the aforementioned overhead and complexities need to be alleviated.

The present chapter investigates the gains brought by dynamic relays in different cellular environments under the presence of ICI. This is achieved with the aid of the versatile Nakagami-m fading distribution (see section 2.2.3). Furthermore some algorithms for expoiting dynamic relays while minimizing the entailed overhead and complexity are devised. In this respect, a novel framework focusing on the downlink is presented, enabling the use of dynamic RSs while minimizing signaling and relay selection complexity. Each MS does not feed back to the BS the channel coefficients between itself and the rest of cell MSs, but only a subset of them determined through the use of a threshold. The considered relaying techniques are the AF and DF ones whose performance is compared in an interference limited environment.

6.2   Signal and System Model

The network consists of B Base Stations with one antenna each and K single antenna MSs per cell uniformly distributed in the cell area. It is assumed that all BSs communicate on the same frequency (full frequency reuse). Downlink communication is taken into account although the presented results are equally valid for the uplink.

6.2.1   Non-relay Assisted Transmission

In the case of direct transmission (non-relay assisted) towards a user i associated with the s-th BS, the received signal is

$$y_i = h_{is}\sqrt{p_s}u_s + \sum_{j=1,j\neq s}^{B} h_{ij}\sqrt{p_j}u_j + n_i \qquad (6.1)$$

where h_{is} corresponds to the channel associated with the s-th BS assigned to MS i, whereas $\sum_{j=1,j\neq s}^{B} h_{ij}\sqrt{p_j}u_j$ corresponds to the ICI and $n \sim \mathcal{NC}\left(0,\sigma^2\right)$ represents the zero mean circularly symmetric AWGN with variance σ^2. u_k, $k = 1,\ldots,B$ represent the unit variance transmit symbols and p_k, $k =$

$1, \ldots, B$ correspond to the transmit power of each BS of the system (equal power allocation is assumed). The instantaneous achievable rate for MS i is

$$C\left(i\right) = \log_2\left(1 + \frac{|h_{is}|^2 p_s}{\sum_{j=1, j\neq s}^{B} |h_{ij}|^2 p_j + \sigma^2}\right). \tag{6.2}$$

The outage probability (OP) for a given source transmit rate $\mathcal{R}$ is

$$P_{\text{out}}^{i,D} = \Pr\left\{C\left(i\right) < \mathcal{R}\right\}. \tag{6.3}$$

The above equation refers to the OP of the direct transmission D (non-relay assisted).

6.2.2 Relay Assisted Transmission

Let $\mathcal{U}_s$ be the set comprising the users of the s-th cell, $|\mathcal{U}_s| = K$. The transmission towards a target user $i \in \mathcal{U}_s$ can be assisted by another user $r \in \mathcal{U}_s \backslash i$, acting as a relay partner. The RSs are assumed to transmit in Half-Duplex (HD) mode, according to which a RS cannot receive and transmit simultaneously. In this case dual-hop transmission requires two time slots. In the first time slot the source transmits symbol u_i. Relay r is in listening mode and receives the signal y_r. If the destination is also in listening mode it receives y_{i1} (case where *diversity* is enabled). In the second time slot the relay node transmits symbol u_r while the source remains silent (orthogonal transmission). The symbol u_r is a function of the relay's received signal y_r and the employed relaying protocol. User i receives signal y_{i2} which is appropriately combined with y_{i1} from the previous time slot. In this paper we assume that Maximal Ratio Combining (MRC) is performed at the destination. Destination i and relay r receive

$$\begin{aligned}
y_{i1} &= h_{is}\sqrt{p_{s1}}u_s + \chi_i \\
y_r &= h_{rs}\sqrt{p_{s1}}u_s + \chi_r \\
y_{i2} &= h_{ir}\alpha_r u_r + h_{is}\sqrt{p_{s2}}u_{s2} + \chi_i
\end{aligned} \tag{6.4}$$

where h_{is}, h_{rs} and h_{ir} are the source-destination, source-relay and relay-destination channel coefficients respectively. It is assumed that the transmit symbols have unit variance and that the channel coefficients remain constant during the 2 time slots of transmission. α_r is an amplification factor which depends on the communication protocol and it ensures that the RS power constraints are met. p_s and p_r represent the power allocation levels of the source and the relay respectively.

It is assumed that in each time slot the total power emanating from a cell is constrained to P. Therefore $p_{s1} \leq P$ and $p_{s2} + p_r \leq P$, where p_{s1} and p_{s2} represent the transmit power of the source node (BS) in the first and second time slot respectively and p_r is the power stemming out of the relay node. In addition

$$\begin{aligned} \chi_i &= z_i + n_i \\ \chi_r &= z_r + n_r. \end{aligned} \tag{6.5}$$

The coefficients z represent the received ICI by the RS and the destination, whereas n is the AWGN; thus χ represents the received interference plus noise (IN). We assume that IN at the destination χ_i remains constant during the 2 time slots of transmission. For exploiting cooperative diversity, several relaying techniques have been proposed, with the most fundamental ones being Amplify-and-Forward (AF) and Decode-and-Forward (DF) [48, 50]. The triangular cooperative model has been taken into account, where there is one source, one relay and one destination (dual-hop communication). According to AF, the RS amplifies the received signal from the source node and forwards it to the destination node without decoding it (*non-regenerative scheme*). The amplification factor is properly adjusted in order for the power constraints of the RS to be met. Nonetheless this relatively simple scheme comes with the detrimental side-effect that the RS apart from the received signal it amplifies its thermal noise together with its received ICI, a factor which limits performance. In DF, the RS decodes and retransmits the received message conditioned that it is able to decode it (*regenerative scheme*). The capacity of both schemes is limited by the source-relay link.

Amplify-and-Forward

With AF the relay node just amplifies its received signal y_r by using an amplification factor α_r which ensures that the relay power constraints p_r are met. Therefore in (6.4), $u_r = y_r$ and

$$\alpha_r = \sqrt{\frac{p_r}{|h_{sr}|^2 \, p_{s1} + \chi_r^2}} \tag{6.6}$$

where $\chi^2 = \mathbb{E}\left[|\chi|^2\right]$. There are two modes of operation for the AF protocol, the orthogonal (OAF) and the non-orthogonal one (NAF). In the case of NAF transmission ($p_{s2} \neq 0$), the equivalent channel matrix is [51]

$$\mathbf{H}_{AF} = \begin{pmatrix} h_{sd}\sqrt{\frac{p_{s1}}{\chi_i^2}} & 0 \\ \frac{h_{rd}\alpha_r h_{sr}\sqrt{p_{s1}}}{\sqrt{|h_{rd}|^2\alpha_r^2\chi_r^2+\chi_i^2}} & h_{sd}\sqrt{\frac{p_{s2}}{|h_{rd}|^2\alpha_r^2\chi_r^2+\chi_i^2}} \end{pmatrix}. \tag{6.7}$$

The achievable rate associated with the equivalent channel matrix is,

$$C_{NAF}(i,r) = \frac{1}{2}\log_2\left(\det\left(\mathbf{I}+\mathbf{H}_{AF}\mathbf{H}_{AF}^H\right)\right). \tag{6.8}$$

The factor $\frac{1}{2}$ is related to fact that transmission takes place in two time slots. In the case of OAF ($p_{s2} = 0$) the equation above reduces to

$$C_{OAF}(i,r) = \frac{1}{2}\log_2\left(1 + \frac{|h_{sd}|^2 p_{s1}}{\chi_i^2} + \frac{|h_{rd}|^2 |h_{sr}|^2 \alpha_r^2 p_{s1}}{|h_{rd}|^2 \alpha_r^2 \chi_r^2 + \chi_i^2}\right). \tag{6.9}$$

Decode-and-Forward

In the case of DF, the relay node fully decodes its received signal, if decoding is possible, and retransmits it to the destination. Therefore $u_r = u_s$ and $\alpha_r = \sqrt{p_r}$ in (6.4). It can operate in two modes like the AF protocol, the orthogonal (ODF) and the non-orthogonal one (NDF). If the signal is decoded correctly, the equivalent channel matrix for NDF transmission is [51]

$$\mathbf{H}_{DF} = \begin{pmatrix} h_{sd}\sqrt{\frac{p_{s1}}{\chi_i^2}} & 0 \\ h_{rd}\sqrt{\frac{p_r}{\chi_i^2}} & h_{sd}\sqrt{\frac{p_{s2}}{\chi_i^2}} \end{pmatrix}. \tag{6.10}$$

Under the DF framework the channel can be seen as a multiple-access channel. The capacity of the DF scheme is limited by the source-relay link, since the relay node needs to correctly decode its received signal. Therefore the following set of constraints need to be met [51]

$$\begin{aligned} R_{t1} &\leq \min\left\{\log_2\left(1 + \frac{|h_{sr}|^2 p_{s1}}{\chi_r^2}\right), \log_2\left(1 + \frac{|h_{rd}|^2 p_r + |h_{sd}|^2 p_{s1}}{\chi_i^2}\right)\right\} \\ R_{t2} &\leq \log_2\left(1 + \frac{|h_{sd}|^2 p_{s2}}{\chi_i^2}\right) \\ R_{max} &\leq \log_2\left(\det\left(\mathbf{I} + \mathbf{H}_{DF}\mathbf{H}_{DF}^H\right)\right) \end{aligned} \tag{6.11}$$

where R_{t1}, R_{t2} refer to the encoding rates of the source during the first and the second time slot respectively. R_{max} refers to the maximum achievable rate of the equivalent multiple-access channel. With respect to (6.11) the capacity of the NDF cooperative protocol when diversity is enabled is

$$
C_{NDF}\left(i,r\right) = \begin{cases} \frac{1}{2} R_{max}, & R_{t1} + R_{t2} \geq R_{max} \\ \frac{1}{2}\left(R_{t1} + R_{t2}\right), & R_{t1} + R_{t2} < R_{max}. \end{cases}
\tag{6.12}
$$

If transmission takes place in an orthogonal manner, the capacity expression reduces to

$$
C_{ODF}\left(i,r\right) \leq \frac{1}{2} \min\left\{ \log_2\left(1 + \frac{|h_{sr}|^2 p_{s1}}{\chi_r^2}\right), \log_2\left(1 + \frac{|h_{rd}|^2 p_r + |h_{sd}|^2 p_{s1}}{\chi_{d2}^2}\right) \right\}.
\tag{6.13}
$$

For all the relaying enabled schemes the end-to-end OP between the source s and the destination user i through the decoding relay r, when source transmits with constant rate $\mathcal{R}$, is

$$
P_{\text{out}}^{r,i} = \Pr\left\{ C\left(i,r\right) < \mathcal{R} \right\}.
\tag{6.14}
$$

6.3   Relay Selection Schemes

Two relay selection schemes are considered, a proactive and a reactive one. In the *proactive* scheme, it is assumed that the source, i.e. the BS, possesses all the relevant system CSI and the relay node is selected before the transmission of the source. The source selects the relay that maximizes mutual information and transmits at a rate equal to this mutual information without any outage probability. Thus, the employed evaluation metric for the proactive scheme with CSI is the maximum attained mutual information (achievable capacity). In the *reactive* scheme, relay selection is performed after the transmission of the source. In the latter case, source transmits at a constant rate and the best relay node is selected amongst the ones that have decoded the source message. As the source lacks CSI before transmission, the OP is our considered metric for this scheme.

6.3.1   Proactive Relay Selection with CSI

We assume that for a specific destination MS i, a set of relay candidates $\mathcal{G} \subseteq \mathcal{U}_s$ is formed. In the limiting case $\mathcal{G}$ comprises all the cell users apart from the destination one. The BS gathers the BS-MS CQI [1] of all relay candidates and the destination and also the CQI between the relay candidates and the destination MS. Therefore the BS can select the best relay node r_b amongst the available ones to aid the transmission towards the i-th MS

$$r_b = \arg \max_{r \in \mathcal{G}, r \neq d} C\left(r, i\right). \tag{6.15}$$

It is possible that the direct transmission from the BS to MS i leads to superior capacity than the transmission through the best relay [78, 79, 101]. Thus the BS compares the capacity of the direct transmission with the one through the best relay and decides whether to use the relay or not

$$C_{final}\left(i\right) = \max\left\{C\left(r_b, i\right), C\left(i\right)\right\}. \tag{6.16}$$

Our considered evaluation metric is the average system capacity

$$\bar{C} = \mathbb{E}\left[C_{final}\left(i\right)\right]. \tag{6.17}$$

The purpose of a smart relay selection algorithm is to minimize the average size of $\mathcal{G}$, as the entailed feedback load (and the selection overhead) is proportional to its cardinality, while avoiding to sacrifice the capacity gains of relaying. We define as the average percentage of cell MSs that are relay candidates per target MS as

$$\bar{P} := \mathbb{E}\left[\frac{|\mathcal{G}|}{|\mathcal{U}_s| - 1}\right] \times 100. \tag{6.18}$$

This percentage corresponds to the average percentage of the total number of CQI indices fed back per cell, representing radio feedback load (signaling overhead). The limiting case of $\bar{P} = 100$ represents the case where all cell users are always relay candidates for a specific target MS ($|\mathcal{G}| = |\mathcal{U}_s| - 1$), thus they all feed back their CQI to the BS (maximum feedback load charge).

---

[1] The CQI metrics considered in this thesis are the channel gain and the average received ICI plus the noise power for each user.

6.3.2   Reactive Relay Selection

In the case of reactive relay selection, the BS transmits at a constant rate $\mathcal{R}$ towards a destination i. The relays that decode the source message form the set $\mathcal{C} \subseteq \mathcal{U}_s$, $i \notin \mathcal{U}_s$. These relays together with the destination feed back to the BS the CSI between them and destination i, so that the relay selection takes place. The best relay partner r_b for destination i, is the one minimizing the end-to-end OP

$$r_b = \arg \max_{r \in \mathcal{C}, r \neq i} C\left(r, i\right) = \arg \min_{r \in \mathcal{C}, r \neq i} P_{\text{out}}^{r,i}. \tag{6.19}$$

Hence, the final OP between the source and destination i is

$$P_{\text{out}}^d = \min \left\{ P_{\text{out}}^{i,D}, P_{\text{out}}^{r_b,i} \right\}. \tag{6.20}$$

Users are considered to be served in a round-robin fashion. The considered evaluation metric for the reactive scheme is the average OP

$$P_{\text{out}} = \mathbb{E}\left[P_{\text{out}}^d\right]. \tag{6.21}$$

It must be noted that this scheme requires that only the relay to destination channel coefficients of the nodes that have decoded the source message (nodes of $\mathcal{C} \subseteq \mathcal{G}$) are fed back to the BS. This feedback overhead is much smaller than the overhead entailed by the proactive scheme. The latter requires that both the BS-relay and the relay-destination coefficients of all the cell users (all cell users are relay candidates) are fed back to the BS.

6.4   Performance over Nakagami-m Fading Channels

This section investigates the performance of dynamic relaying in different cellular environments prone to ICI, so that to apprehend under which fading conditions relaying is more beneficial. As detailed above, cellular systems employing the triangular relaying model are considered, and for a specific target user at most one relay partner is selected. The considered performance metrics are the average achievable rate of the system or the outage probability of the system depending on whether the employed relay selection scheme is a *proactive* or a *reactive* one. It should be pointed out that in this section all cell users are considered as relay candidates (full overhead). Our purpose is to show how the benefits of dynamic relaying vary over different fading environments; thus limiting the overhead related with relay selection is beyond the scope of this evaluation (for this purpose see section 6.5).

Different cellular environments are modeled by adjusting the m parameter of the Nakagami-m^2 distribution as this parameter determines the severity of small-scale fading and the degree of LOS. Interestingly, it turns out that the more severe the fading is, the greater the gains of proactive relaying are. Thus, dynamic relaying is a promising solution for boosting the achievable rate in systems plagued by fading.

6.4.1   Performance Evaluation

A two-tier cell network (19 BSs overall with radius of 2 km) has been considered with the central cell being our cell of interest as it captures well the effect of ICI. Both MSs and BSs are assumed to have single antennas. The channel coefficient between the BS and MS antennas is given by (2.19) of section 2.7. The envelope of the small-scale fading $|\Gamma_{k,\ell}|$ is given by the Nakagami-m distribution of (2.10). Shadowing is not included and path-loss follows the model of (2.4). BS antennas are assumed to be omnidirectional with a 9 dB gain on the elevation. Transmit power is determined by the System SNR, explained in section 2.7.

Various fading conditions have been assumed modeling different macrocellular and microcellular environments with the aid of Nakagami-m fading distribution. The plots have been drawn for ODF relaying and for a System SNR of 20 dB (ICI limited regime).

In Fig. 6.1, $\overline{C}$ is plotted versus the number of cell users K for the *proactive* relaying scheme and for two different types of fading environments, fading with NLOS and fading with LOS. For the former case, three different NLOS fading environments are considered:

1. A bad urban environment, where BS-MS and MS-MS channels as well as ICI are subject to Rayleigh fading, i.e. $m_D, m_{ICI} = 1$ [3].

2. A macrocellular one, where BS-MS (D channels) and MS-MS channels as well as ICI are subject to Nakagami-m fading with $0.5 \leq m_D, m_{ICI} \leq 1$.

---

[2]The properties of this distribution are outlined in section 2.2.3.

[3]m_D denotes the m parameter of the useful BS-MS channels whereas m_{ICI} denotes the m parameter of the interfering channels.

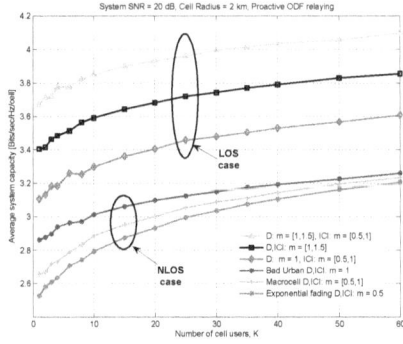

Figure 6.1: System capacity versus the number of cell users (ODF proactive relaying, System SNR = 20 dB) under different fading conditions.

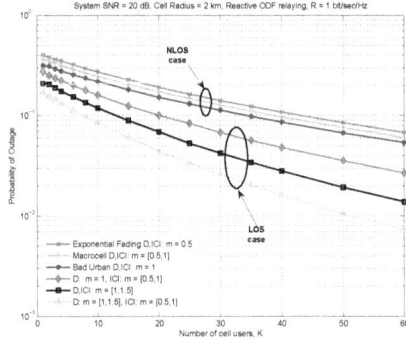

Figure 6.2: Outage probability versus the number of cell users (ODF reactive relaying, System SNR = 20 dB, source rate $\mathcal{R} = 1$ bis/sec/Hz) under different fading conditions.

3. An environment with severe fading, where BS-MS and MS-MS channels as well as ICI channels experience exponential fading, i.e. $m_D, m_{\mathrm{ICI}} =$

0.5.

As depicted in this figure, for all NLOS fading conditions under consideration, $\overline{C}$ increases as K increases and for large K all $\overline{C}$ curves converge. Obviously, increasing K increases the options for relay selection. It must be noted that for $K = 1$, there exists no relay to be selected and thus it is the lower bound for the average system capacity. More importantly, it is shown that relay-assisted transmission results in larger increase on $\overline{C}$ for increasing K as fading conditions become more severe. Furthermore, Fig. 6.1 illustrates the capacity performance in various LOS conditions for the BS-MS and/or MS-MS and/or ICI channels. For all LOS conditions under consideration for the BS-MS and MS-MS channels, increasing K results in similar improvements on $\overline{C}$. Moreover, the resulting $\overline{C}$ becomes larger as LOS conditions become stronger.

Under the aforementioned assumptions for the fading conditions, Fig. 6.2 illustrates the OP performance of the *reactive* relaying scheme for the case that the source transmits with a constant rate of $\mathcal{R} = 1$ bits/sec/Hz. In this case, the trend observed in Fig. 6.1 is reversed; the gain in OP becomes larger as LOS gets stronger whereas for different NLOS conditions the gain in OP remains similar. In Fig. 6.3, the capacity performance of the proactive scheme for various LOS conditions for the MS-MS channels and NLOS ones for all BS-MS and ICI channels is presented. Clearly, stronger LOS conditions for the MS-MS channels results in larger $\overline{C}$ for increasing K.

Generally we can conclude that:

1. For all fading conditions under consideration, relay-assisted transmission becomes more efficient as the number of relay candidates increases.

2. Severe fading can be efficiently mitigated by employing dynamic relays with the use of a proactive relay selection scheme with CSI.

3. If full CSI is not available and a reactive scheme is employed, relaying is more beneficial in the LOS regime as its gains increase proportionally to the degree of LOS.

6.5   Limiting the Overhead of Proactive Relaying with CSI

Proactive relaying with CSI is the most intelligent way of relaying as system capabilities are exploited to the maximum. However it entails a heavy

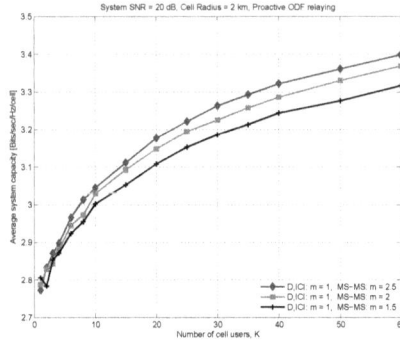

Figure 6.3: System capacity versus the number of cell users (ODF proactive relaying, System SNR = 20 dB). MS-MS channels are assumed to be LOS ones.

feedback overhead, as the source (the BS in the downlink) needs to gather all BS-MS and MS-MS CSI in order to make the decision upon relaying. Therefore it is critical that this heavy load is alleviated. In this section the relaying patterns in the cell are observed under a full feedback overhead framework, and this information is exploited in order to effectively reduce this overhead. These observations are coupled with some novel ideas that allow the system to benefit from dynamic relaying while the overhead remains affordable.

Throughout this section MSs are assumed to be served in a round-robin manner in order to guarantee fairness, and transmission towards a target MS can be aided by another MS of the cell acting as a relay partner.

6.5.1   Relay Selection with Full Overhead

In the case entailing full CSI feedback overhead, the BS gathers all inter-user and user-BS CSI and selects for a specific destination i its best relay partner r_b amongst the users of the cell. A proactive communication protocol adapting [53] to the cellular case, which permits the exploitation of dynamic relays while decisions are made by the BS, is given by Algorithm 5. The CQI considered is the inter-user and user-BS channel gain and the interference plus noise power of both the potential RSs and the destination. The potential relay partners for the i-th MS are the rest of cell users, $\mathcal{G} = \mathcal{U}_s \backslash i$, which entails maximal feedback overhead. The BS gathers all the necessary CQI, and it can transmit at a rate equal to the capacity (which is a function of the employed scheme) without any probability of outage. Under this framework all cell users can profit from relaying, although as it can be seen in section 6.5.1 the probability that a destination MS chooses a relay partner or not highly depends on the MS position due to the effects of BS antenna gain, pathloss and ICI. Consequently this dependence can be leveraged in order to greatly reduce the overhead entailed by the relay selection process (see section 6.5.2).

Algorithm 5 Relay selection with full overhead

1: *The BS s transmits a RTS (ready-to-send) packet to target node i.*
2: Each MS $k \in \mathcal{U}_s$ overhears the RTS and estimates the channel gain to its BS $|h_{ks}|^2$ and its INP.
3: *Node i transmits a CTS (clear-to-send) packet*
4: Each MS $r \in \mathcal{U}_s \backslash i$ overhears the CTS and estimates its channel gain to node i, $|h_{ir}|^2$.
5: Each MS $r \in \mathcal{U}_s \backslash i$ feeds back to the BS its channel gain to destination $|h_{ir}|^2$ and its INP χ_r^2.
6: Target node i feeds back its channel gain to the BS $|h_{is}|^2$ and its INP χ_i^2.
7: *Relay selection by the BS:* $r_b = \arg \max\limits_{r \in \mathcal{U}_s \backslash i} C(r, i)$.
8: **if** $C(r_b, i) > C(i)$ **then**
9:    Node r_b is employed, $C_{final}(i) = C(r_b, i)$
10: **else**
11:    BS transmits directly to node i, $C_{final}(i) = C(i)$
12: **end if**

Observing Relaying Patterns

The algorithm presented above entails very high feedback overhead. Consequently it is of great interest to observe the relaying patterns in a cell in order to devise schemes that achieve similar performance with much lower feedback charge and complexity, by disabling relaying for MSs that do not profit from it. Therefore numerical experiments are conducted in a 19 cell network (cell radius is 1 km). The central cell is our cell of interest since it captures all the ICI originating in the other cells. The channel coeffi-

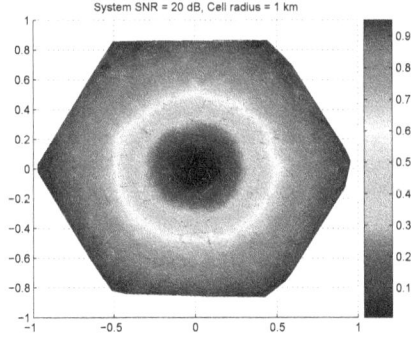

Figure 6.4: Probability of relay partner choice P_{rel} as a function of the MS target position when DF is enabled (Proactive ODF).

cient between the i-th MS and the j-th sector is given by (2.19) of section 2.7. NLOS channels are assumed, thus multipath fading is described by the Rayleigh distribution of (2.6). Shadowing follows the log-normal distribution of (2.5) with standard deviation 8 dB and path-loss follows the model of (2.4). BSs are assumed to have omnidirectional antennas with a 9 dB gain on the elevation. Transmit power is determined by the System SNR.

An indicative metric of the relaying patterns in a cell is the probability $P_{rel}(i)$ that the utilization of the best relay partner r_b of MS i leads to superior capacity that the direct transmission from the BS to i. This probability is

$$P_{rel}(i) = \Pr\{C(r_b, i) > C(i)\}. \tag{6.22}$$

$P_{rel}(i)$ as a function of the target MS position in a densely populated cell and in the ICI limited regime (System SNR = 20 dB) can be seen in Fig. 6.4. Notably, the further away a target MS is from the BS, the more likely transmission via its best relay is chosen as it experiences high pathloss to the BS. Furthermore the antenna gain of the BS provides a strong direct BS-MS channel to MSs closely located to the BS which consequently are unlikely to choose a dual-hop transmission through a relay partner. In addition the MSs located close to the cell edge experience high ICI and they are likely to choose an ICI resilient relay partner for enhancing their capacity. $P_{rel}(i)$ is

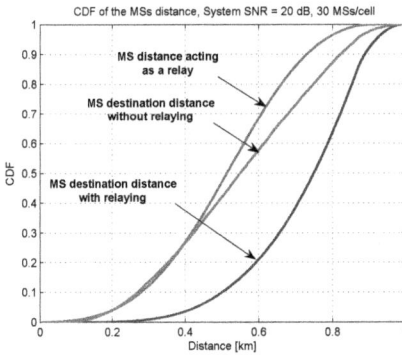

Figure 6.5: CDF of the MS target distance which chooses relaying, the MS target distance which does not choose relaying and the RS distance (30 MSs/cell, ODF, cell radius = 1 km).

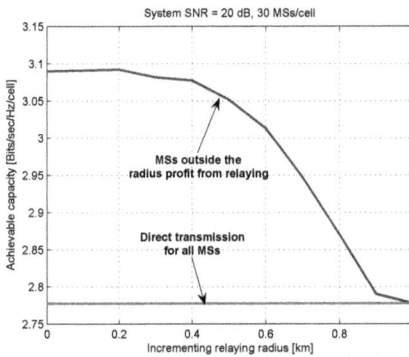

Figure 6.6: System capacity versus an increasing relaying radius (30 MSs/cell, System SNR = 20 dB, ODF)

significantly lower in a cell with much less users comparing to the scenario depicted in Fig. 6.4, as the more the available users are the higher the probability is that a good relay partner is found. This can be seen in Fig. 6.5 where it is plotted the empirical CDF of the distance from the BS of the MS targets preferring a relay partner, the RS distance and the distance of the MS targets preferring direct transmission from the BS for 30 MSs/cell (distances are measured from the cell centre).

Another experiment leading to similar conclusions is to permit profiting from relaying only target MSs located outside a prespecified radius. The MSs whose distance is less than this radius receive their signal directly from the BS. In Fig. 6.6 the average achievable capacity $\bar{C}$ is plotted as a function of an increasing relaying radius for 30 MSs/cell and System SNR of 20 dB. When radius is 0 km all users can profit from relaying (capacity achieves its maximum point) whereas when radius is 1 km (considered cell radius), all users are served by direct transmission from the BS (minimum value for capacity). It is noticeable that for radii less than 0.5 km the capacity does not dimish significantly from its maximum value; this means that for 30 MSs/cell and a BS antenna with 9 dB gain, relaying should not be enabled for users at a distance greater than half of the cell radius. Therefore the incurred feedback overhead for relay selection can be substantially alleviated if relaying is completely switched off for MSs relatively close to the BS.

6.5.2   Proactive Relay Selection with Limited Overhead

A simple way of taking the observations of the previous section into account for reducing the feedback overhead associated with relaying, is that only users whose distance from the BS exceeds a threshold d_{t1} are allowed to request the aid of a potential relay partner. In addition, for the MSs that request the aid of a relay node, it is much more likely to find a good relay partner closer to them than to the BS. This is due to the antenna gain at the BS which renders the BS-relay link stronger on average than the relay-MS link as the relay node is in principle located between the BS and the target MS. Consequently we consider as potential relay nodes for a target MS, the nodes whose distances to the target are below a second threshold d_{t2} (these MSs are inside a circle whose centre is the target node, see Fig. 6.7). Inside this conceptual circle, there can be nodes which are further away from the BS than the target; these nodes are not likely to provide gains acting as relays since they experience greater attenuation to the BS than the target. Thus we can consider as potential relay partners the nodes inside this circle which are closer to the BS than the target by applying a third threshold

d_{is}. If MSs know their distance (or attenuation) to the BS and to the target MS, the abovementioned observations can be taken into account in devising the relay selection protocol given by Algorithm 6. This framework leads to significant reductions in signaling overhead comparing to Algorithm 5 without compromising performance; this is because only a very limited but suitable subset of cell users become relay candidates.

Algorithm 6 Relay selection with limited overhead

Require: Define distance thresholds d_{t1}, d_{t2}
  1: *The BS s transmits a RTS packet to target node i*
  2: Each MS $k \in \mathcal{U}_s$ overhears the RTS and estimates its channel gain to the BS $|h_{ks}|^2$ and its INP.
  3: **if** $d_{is} > d_{t1}$ **then**
  4:    *Node i transmits a CTS/ON packet containing its distance to the BS d_{is}*
  5:    Each MS $r \in \mathcal{U}_s \backslash i$ overhears the CTS/ON and estimates its channel gain to target $|h_{ir}|^2$ and its INP.
  6:    **for** each MS $r \in \mathcal{U}_s \backslash i$ **do**
  7:       **if** $d_{ir} \leq d_{t2}$ and $d_{sr} < d_{is}$ **then**
  8:          MS r becomes relay candidate, $r \in \mathcal{G}$
  9:          MS r feeds back to the BS its channel gain to target $|h_{ir}|^2$ and its INP χ_i^2.
 10:       **end if**
 11:    **end for**
 12:    Target node i feeds back its channel gain to the BS $|h_{is}|^2$ and its INP χ_i^2.
 13:    *Relay selection by the BS: $r_b = \arg\max\limits_{r \in \mathcal{G}} C\left(r, i\right)$.*
 14:    **if** $C\left(r_b, i\right) > C\left(i\right)$ **then**
 15:       Node r_b is employed, $C_{final}\left(i\right) = C\left(r_b, i\right)$
 16:    **else**
 17:       BS transmits directly to node i, $C_{final}\left(i\right) = C\left(i\right)$
 18:    **end if**
 19: **else**
 20:    *Node i transmits a CTS/OFF packet*
 21:    BS transmits directly to node i, $C_{final}\left(i\right) = C\left(i\right)$
 22: **end if**

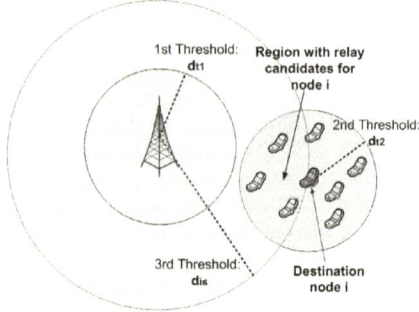

Figure 6.7: The proposed relay selection technique (Algorithm 6). 3 thresholds are employed for determining the relay candidates for target node i: d_{t1}, d_{t2} and d_{is}.

6.5.3   Performance Evaluation

For the numerical evaluation of the proposed scheme, the channel and system model described in section 6.5.1 has been employed. In Fig. 6.8 it is plotted the average percentage $\bar{P}$ of cell users considered as relay candidates per target MS (6.18) as a function of the distance threshold d_{t2} for different versions of Algorithm 6 and for 30 MSs/cell. The solid red curve refers to the case where only threshold d_{t2} is applied for determining relay candidates whereas the dashed red curve represents the case where both d_{t2} and d_{t1} thresholds are applied (d_{t1} is set to 0.5 km). For the blue curves d_{t2} and d_{is} thresholds are applied. More specifically the dashed blue curve corresponds to the case where relaying is enabled only for users whose distance from the BS is grater than $d_{t1} = 0.5$ km and this attains the minimum overhead charge (all 3 thresholds are applied). It can be seen that for d_{t2} set to 0.5 km and when all thresholds are applied (d_{t1}, d_{t2} and d_{is}) the system is charged only with 10 per cent of the potential feedback overhead (lowermost curve). It should be noted that the protocol described by Algorithm 5 always charges the system with 100 per cent overhead.

In figure 6.9 it is plotted the average achievable rate performance for the proposed overhead reduction scheme against threshold d_{t2} (only threshold d_{t2} is applied) for the two modes of DF and AF protocols with enabled diversity (destination node listens in both time slots) in order to investigate which

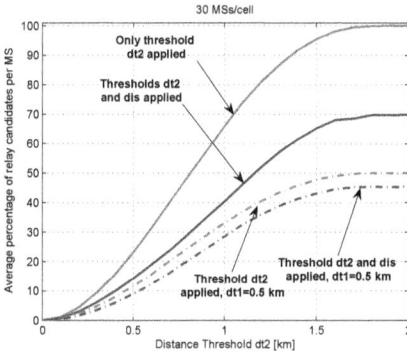

Figure 6.8: Average percentage of relay candidates per target MS versus the distance threshold d_{t2} for different versions of Algorithm 6 (threshold d_{t2} is always applied).

of the considered relaying protocols [4] performs the best. In each time slot the power emanating from each cell is constrained. Consequently BS and RS share the available cell power during the second time slot (non-orthogonal schemes). It can be seen that the performance of all the considered schemes saturates when the threshold distance reaches 0.5 km (maximum performance is achieved). DF schemes outperform AF ones. More specifically, the NDF scheme achieves the greatest sum-rate performance, although the ODF performs the same when the threshold distance is greater than 0.6 km. Regarding the AF schemes, the OAF outperforms the NAF. Notably NDF's performance coincides with the one of ODF for $d_{t2} > 0.5$ km. ODF and OAF are the chosen protocols for fully evaluating the proposed protocol of Algorithm 6, as they are much simpler than NDF and NAF and lead to very similar performance (especially ODF).

In Fig. 6.10 it is plotted the average capacity $\bar{C}$ (6.17) against threshold d_{t2} for different versions of Algorithm 6 (threshold d_{t2} is always applied) and for both DF and AF relaying (System SNR is set to 20 dB). For all the considered versions of Algorithm 6, performance saturates approximately when threshold d_{t2} reaches 0.5 km (maximum performance is achieved); it can be inferred that good relay candidates for a target MS are located no

---

[4]The considered relaying protocols are the ODF, NDF, OAF and NAF.

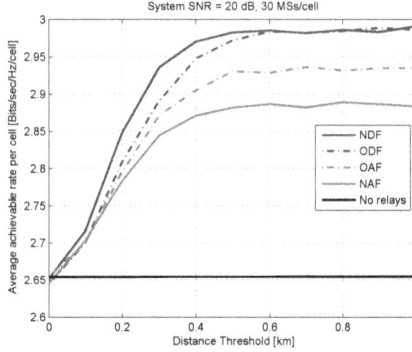

Figure 6.9: Sum-rate versus distance threshold for the proposed framework.

further than 0.5 km from it. The red curve represents the case where only threshold d_{t2} is applied, and this achieves the best performance. When $d_{t2} >$ 0.5 km, the saturated red curve also represents the capacity performance of Algorithm 5 entailing full feedback overhead; in this regime the 2 frameworks coincide although Algorithm 6 implies significantly less feedback. It can be also seen that applying all the 3 thresholds (d_{t2}, d_{is} and d_{t1} - green curve) results in a small performance degradation comparing with the case where only d_{t2} is applied (red curve) while it provides substantial feedback load alleviation.

With $d_{t2} = d_{t1} = 0.5$ km and when d_{is} is applied (green curve), only 10 per cent of the total available users are considered as relay candidates per destination node on average, $\bar{P} = 10$ (see lowermost blue curve of Fig. 6.8). For 30 MSs/cell this translates to only 3 relay candidates per MS target on average.

6.6   Conclusions

Although the importance of cooperative diversity has been well recognized, utilization of dynamic relay nodes in cellular systems remains challenging due to the entailed high signaling load and complexity. It is crucial that relay partner selection is done in an opportunistic way in order to optimize network performance while keeping overheads to a minimum. Furthermore

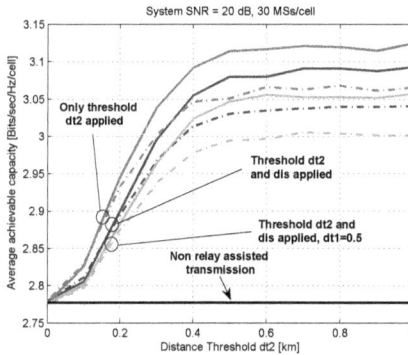

Figure 6.10: Sum-rate versus distance threshold d_{t2} for different versions of Algorithm 6 (threshold d_{t2} is always applied). Solid lines: DF, Dashed lines: AF.

it is important to assess under which fading conditions relaying is more beneficial. In this chapter we have assessed the gains of relaying in different cellular environments with the aid of Nakagami-m fading distribution and we have presented a novel framework for exploiting dynamic relays in a full frequency reuse network. According to this framework, for serving a specific MS of the cell, not all the cell MSs should be considered as potential relay partners. It is sufficient that a small subset of the overall MSs, the ones close to the destination node (determined by a distance threshold) become RS candidates. In this fashion, close to maximum performance can be attained while feedback overhead and complexity are dramatically reduced, bringing relay utilization closer to practice.

Chapter 7

Other Contributions

Most of the previous chapters presented ways of exploiting Multicell-MIMO techniques while limiting their overhead in order to render them practically feasible. The basic principle underpinning Multicell-MIMO is node collaboration; some nodes of the network (in this case the BSs) exchange information and behave as a virtual antenna array. The same basic concept applies to the cooperative relaying techniques presented in the previous chapter. In this chapter, inspired by the same concept, it is proposed that the actuators of a Sensor Actuator Network (SANET) cooperate in order to guarantee that sensing is more efficient.

SANETs refer to a group of sensors and actuators wirelessly linked to perform distributed sensing and actuation tasks [102, 103]. In such networks, sensors gather information about the physical world while actuators are usually resource-rich devices with higher processing and transmission capabilities, and longer battery life. Actuators collect and process sensor data and perform *actions* on the environment based on the gathered information.

Unlike wireless sensor networks where the central entity, i.e. the sink, performs the functions of data collection and coordination, in SANETs new networking phenomena called *sensor-actuator* and *actuator-actuator* coordination may occur. In particular, sensor-actuator coordination provides the transmission of event features from sensors to actuators. After receiving event information, actuators may need to coordinate with each other in order to make decisions on the most appropriate way to perform the actions.

Each sensor node is associated with an actuator which is the destination
of the sensor data. Actuators receive sensor data in a multi-hop fashion
and transmit the scheduling information to them in a single hop fashion.
In order to prevent sensor data collisions, actuators transmit *time schedules*
which coordinate sensor multi-hop transmission. Therefore each sensor after
receiving the scheduling information from its associated actuator transmits
its data in the right time slot.

However due to the impairments introduced by the wireless channel (effects of pathloss and fading), it is possible that some sensor nodes, more
likely the ones that are distant from the actuator, would not be able to
decode their scheduling information. This is because some sensor nodes
would probably receive the signal containing scheduling information at a
very low SNR. Consequently they will remain *inactive* (they will not know
when to transmit), a fact that could create some inactive zones in the sensing field. This would result to incomplete information reception, a situation
that needs to be overcome for the uniform monitoring of the sensing field.
A potential solution to this would be the use of positive and/or negative
acknowledgments (ACKs and/or NACKs) with respect to the reception of
scheduling information. In this fashion, for the sensor nodes that cannot
decode their scheduling information, multi-hop transmission of their schedules can be employed. However this would result in a significant overhead
burden in terms of time and energy waste of the sensor nodes, that can
reduce their lifetime. Furthermore that type of solution would increase the
complexity of the employed protocols.

The problem of sensor inactivity can be effectively faced on the physical
layer without increasing the protocol complexity and dissipating extra energy from sensor nodes. Actuators can cooperate and form a distributed antenna array, a concept that has been proposed for cellular communications
(see previous chapters). The array jointly performs adaptive beamforming and distributes the time schedule to each sensor node. Sensors receive
scheduling information at much higher power due to the array gain that results from beamforming and to the exploitation of *macro-diversity* which is
inherent to the distributed nature of a SANET. This results in a significant
reduction in the number of inactive sensors for a given transmit power level.
The cost is the need of CSIT.

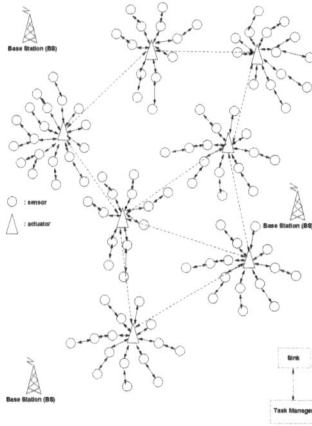

Figure 7.1: Architecture of SANETs.

7.1 Signal and Network Model

A static wireless sensor-actuator network with N sensor nodes, M actuators nodes is considered as shown in Fig. 7.1. Each sensor and actuator is equipped with an omnidirectional antenna. Actuators are inter-connected via a backhaul network (wireline or wireless). It is assumed that an equal number of sensors K is assigned to each actuator, so as $M \times K = N$. A sensor node can decode a transmission from a neighboring sensor successfully if the experienced SNR or SINR is above a certain threshold.

The channel between the ith sensor node and the jth actuator is given by (2.19). NLOS channels are assumed, thus multipath fading is described by the Rayleigh distribution of (2.6). Shadowing follows the log-normal distribution of (2.5) with standard deviation 8 dB. The pathloss constant and exponent are chosen according to the COST-231 model, where actuator height is assumed to be $10\,m$ and sensor node height $10\,cm$.

7.2 Actuator to Sensor Transmission Schemes

In this section, three different actuator-to-sensor transmission schemes are presented together with their analysis and performance evaluation. Actuators can all transmit at the same frequency and therefore interfere with each other. They can also transmit at different frequencies in order to avoid interfering with each other at the cost of employing a higher frequency reuse factor.

7.2.1   Transmission at a Single Frequency - Reuse Factor 1

In this case, each actuator communicates with the sensor nodes that are assigned to it. The actuator broadcasts a packet containing scheduling information, for the sensors nodes attached to it, at the same frequency. Each sensor node receives together with useful scheduling information, Co-Channel Interference (CCI) from other actuators. The received signal of the sensor node i is

$$y_i = h_{ij}\sqrt{p_j}x_j + \sum_{k \neq j} h_{ik}\sqrt{p_k}x_k + n \qquad (7.1)$$

where $i = 1, 2, \ldots, K$, j is the actuator that the sensor i is assigned to, p_n for $n = 1, 2, \ldots, M$ is the transmit power of the m-th actuator, n is the AWGN component with variance σ^2, and x_n for $n = 1, 2, \ldots, M$ is the unit variance transmitted scheduling information of the n-th actuator. Throughout this chapter it is assumed that all actuators transmit on the same power level.

The packet x_j contains the schedules of all sensor nodes attached to actuator j. $\sum_{k \neq j} h_{ik}\sqrt{P_k}x_k$ represents the detrimental CCI term. Therefore, the SINR of the i-th sensor node is

$$\gamma_i = \frac{|h_{ij}|^2 p_j}{\sum_{k \neq j} |h_{ij}|^2 p_k + \sigma^2}. \qquad (7.2)$$

If γ_i is below a certain threshold T ($\gamma_i < T$), sensor node i is unable to decode its scheduling information and therefore it is unable to resolve when to transmit its sensed data. Thus, it will remain *inactive*.

The advantage of this scheme is that each actuator, in order to distribute sensor scheduling information, broadcasts a packet that contains all sensor schedules. Therefore, in one time slot, all schedules are distributed. However, each sensor needs to go through all the contents of the scheduling packet in order to find its own schedule, a fact that increases decoding complexity. The main disadvantage is that some sensor nodes might remain inactive as described above.

7.2.2   Transmission at Different Frequencies - Higher Reuse Factor

In this case also, each actuator communicates with the sensor nodes that are associated with it. Each actuator broadcasts its scheduling information at a different frequency. This eliminates CCI at the cost of employing a higher frequency reuse factor. The received signal at the sensor i is then

$$y_i = h_{ij}\sqrt{p_j}x_j + n \tag{7.3}$$

where $i = 1, 2, \ldots, K$. The SNR of the i-th sensor node is

$$\gamma_i = \frac{|h_{ij}|^2 p_j}{\sigma^2}. \tag{7.4}$$

The advantage of this scheme comparing to the frequency reuse factor 1 is the elimination of CCI. CCI degrades the received SNR and therefore increases the probability of sensor inactivity. By using different frequencies for each actuator, the number of inactive sensors is decreased for a given level of transmit power.

7.2.3   Actuator Cooperation - Joint Beamforming

This scenario has been inspired by MCP techniques proposed in the previous chapter; actuators are assumed to be interconnected via high speed backhaul links (wireline or wireless). After an initial handshake between a sensor and its associated actuator (min. hop fashion, more details on this assignment are provided in [103]), each actuator transmits a training sequence. Then each sensor estimates the channel between itself and all the actuators, and it transmits this set of channel coefficients to its associated actuator in a multi-hop fashion. Therefore CSIT is obtained. Furthermore, each actuator determines the schedules for its associated sensors. Actuators exchange their local CSIT and their scheduling information via the backhaul links, and jointly perform MRC beamforming in order to transmit the scheduling information to each sensor. Hence, actuators form a distributed antenna array. The transmission of the scheduling information is done in a round robin fashion and at the same frequency. Each sensor has a channel vector $\mathbf{h}_i = [h_{i1}, h_{i2}, ..., h_{iM}]^T$. In order for the per-actuator power constraint to be satisfied, each actuator j transmits to sensor i

$$A_{ij} = \frac{h_{ij}^*}{|h_{ij}|}\sqrt{p_j}s_i. \tag{7.5}$$

The received signal of the sensor node i is then

$$
\begin{aligned}
y_i &= \sum_{j=1}^{M} h_{ij} A_{ij} + n = \\
&= \sum_{j=1}^{M} |h_{ij}| \sqrt{p_j} s_i + n
\end{aligned}
\tag{7.6}
$$

where $i = 1, 2, ..., N$ and s_i is the unit variance schedule assigned to sensor node i. Thus the SNR of the sensor node i in the case of equal power allocation $p_j = p$ for $j = 1, \ldots, M$ is

$$
\gamma_i = \frac{p \left(\sum_{j=1}^{M} |h_{ij}| \right)^2}{\sigma^2}.
\tag{7.7}
$$

Joint beamforming enhances the received SNR due to the array gain and the exploitation of macro-diversity which is inherent in a SANET. Therefore, this scheme provides a robust way of minimizing sensor inactivity. This is achieved at the cost of CSIT at the actuators. Furthermore, multiple time slots are needed in order to deliver the schedule to all sensor nodes, since actuators transmit to one sensor node at a time.

7.3   Performance Evaluation

The performance of the schemes presented above is evaluated in terms of the number of inactive sensors that results from each transmission scheme. A number of sensors is deployed uniformly in a hexagon with a radius of 1 km. Three actuators are assumed at the three vertices of the hexagon separated by an angle of 120. Actuator antennas are consider to have a gain of 12 dB (gain on the elevation), whereas sensor node antennas have a gain of 1 dB. Through Monte-Carlo simulation, the average number of inactive sensors is calculated for each transmission scheme as a function of the actuator transmit power. Averaging is performed over sensor node positions and channel realizations. A sensor is assumed to be inactive if its received SNIR or SNR (depending on the employed scheme) is below the threshold of 1 Watt.

In Fig. 7.2 it is plotted the average number of inactive sensors versus the actuator transmit power for 1200 deployed sensors. It can be seen that for the power of -12 dBw, inactive sensor zones are almost completely eliminated

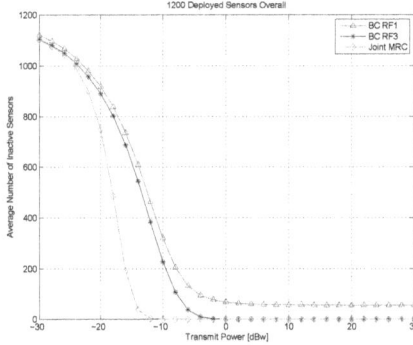

Figure 7.2: Average Number of Inactive Sensors vs. TransmitPower.

in the case of MRC beamforming. In the case of Reuse Factor 3, inactive zones are eliminated when the transmit power is approximately 0 dBw and in the case of Reuse Factor 1 the average number of inactive sensors saturates approximately at 0 dBw.

In Fig. 7.3 it is plotted the average number of inactive sensors against the total number of deployed sensors for a different number of deployed sensor nodes, when actuators transmit power is -12 dBw. It can be clearly seen that the joint MRC beamforming scheme outperforms the simple Reuse 3 broadcasting, as the average number of inactive sensors is almost 0 for that power level.

In figures 7.4, 7.5 and 7.6, the probability of inactivity can be seen in the different areas of the hexagon for the three different transmission schemes considered, when actuators transmit at power equal to -12 dBw. In the cases of reuse 1 and reuse 3 schedule broadcasting, the centre of the topology experiences a significant probability of inactivity. In a real system implementation, this would result in an important loss of information. On the contrary, joint beamforming almost eliminates inactive areas in the sensing field at this power level. This turns out to be a very effective actuator transmission scheme that greatly reduces the amount of transmit power needed to ensure very low sensor inactivity. This is because of the beamforming SNR gains and the macro-diversity gains that are provided by the spatially distributed transmitting actuators.

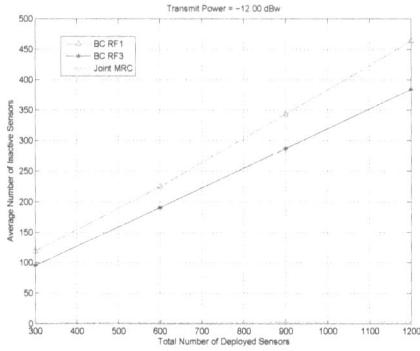

Figure 7.3: Average Number of Inactive Sensors Vs. Total Number of Deployed Sensors.

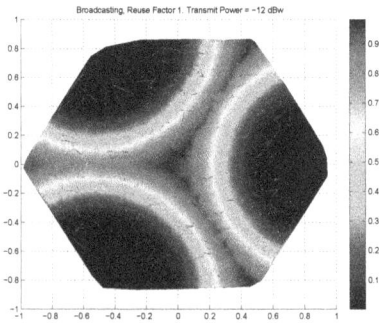

Figure 7.4: Probability of Sensor Inactivity in the areas of the sensing field for the case of Reuse Factor 1 Schedule Broadcast Transmission.

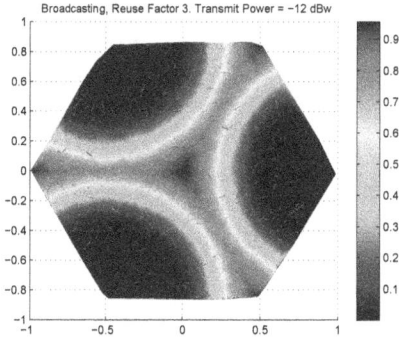

Figure 7.5: Probability of Sensor Inactivity in the areas of the sensing field for the case of Reuse Factor 3 Schedule Broadcast Transmission.

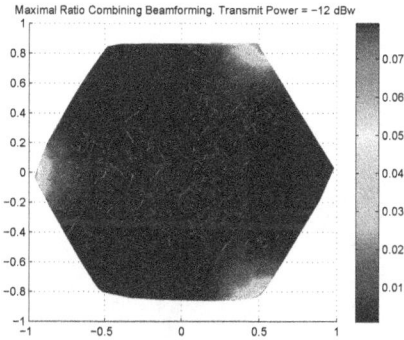

Figure 7.6: Probability of Sensor Inactivity in the areas of the sensing field for the case of joint Maximal Ratio Combining Beamforming.

7.4   Conclusions

This chapter proposes that some techniques inspired from Multicell-MIMO
and generally from cooperative communications are applied to SANETs.
More specifically it addresses the problem of inactive regions in a sensing
field by letting actuators exchange their CSIT and jointly perform beamforming in order to deliver scheduling information to sensor nodes. The
gains of cooperation have been shown by simulating the average number of
inactive sensors for the case of single actuator transmission and cooperative
transmission. This turns out to be a solution that can better exploit sensors
and provide significantly improved coverage of the sensing field.

Chapter 8

Conclusion

The main subject of this thesis is Multicell-MIMO (or MCP) and techniques that can help render this promising technology practically applicable. In Multicell-MIMO enabled systems, a number of BSs behave as a distributed MIMO array and this can efficiently mitigate ICI. This has been shown to significantly enhance throughput and fairness of wireless systems plagued by ICI and it has been identified as a key concept underpinning future 4G systems. However Multicell-MIMO comes with some drawbacks that need to be addressed in order for it to be brought into practice. These drawbacks are related with the overheads entailed by MCP, namely signaling overheads (CSI feedback) and architectural overheads (Control Unit, backhaul links). The present thesis attempts to address these overheads. MCP can bring enormous gains if all the BSs of the network form a single cooperation cluster and behave as a huge distributed antenna array. In this case all ICI can be mitigated and communication is only noise limited. However the overheads entailed by this ideal conception are prohibitive. In this thesis it is proposed that network BSs are grouped into clusters of limited size (each cooperation cluster contains a small number of BSs) as both signaling and infrastructural overheads are proportional to the cluster size. As a result, the way of forming clusters becomes critical for MCP. More specifically we have proposed that limited BS cooperation clusters can be formed in a dynamic manner, taking into account signal fading conditions. This has been shown to be an effective means of BS clustering greatly outperforming static

107

clustering solutions (cases where only neighboring BSs form clusters).

A crucial overhead of MCP systems is the one related with its architectural conception. According to the typical conception for the architecture of MCP enabled systems, the BSs of each cooperation cluster are interconnected via low latency backhaul links. Furthermore a Central Unit is the entity responsible for gathering user CSI of the cooperation cluster, performing user scheduling and designing the parameters of transmission. In the downlink under FDD, this architectural conception is unavoidable if users feed back their CSI to one BS, as it is done in conventional cellular systems. In this dissertation it has been proposed that users communicate their CSI not only to one BS but to all cooperating BSs (broadcast feedback). In this fashion BSs do not need to exchange CSI through the backhaul links as they already possess global CSI of their cluster. Moreover the use of a central Control Unit is unnecessary as scheduling and transmission operations can be performed in a fully distributed way by the cooperating BSs.

Another major issue impeding the deployment of Multicell-MIMO systems, is the very high feedback and backhaul overhead that is required. As in the conventional multiuser MIMO systems operating under FDD, in MCP networks users need to estimate and feed back the channel coefficients related to at least all cooperating antennas (antennas of all the cluster BSs) in order for user scheduling and transmission to take place. Therefore the feedback load in MCP enabled networks is far greater than the one of monocellular multiuser MIMO systems. In the present dissertation we propose a technique coined *selective feedback* according to which only channel coefficients whose channel gain exceeds a given threshold are fed back by the users to the network infrastructure. This feedback load reduction can be exploited in reducing backhaul overhead through scheduling. This approach is shown to achieve a good tradeoff between performance and overhead, making MCP practically feasible.

Apart from Multicell-MIMO, another promising technology that can increase spectral efficiency and fairness of wireless systems is cooperative relaying. Cooperative relaying entails that some relay nodes receive the source message and retransmit it to its destination. They can be components of the system infrastructure placed in specific positions of a cell (static relays) or users of the system that relay messages intended for other users (dynamic relays). In this thesis we have investigated the application of dynamic relays which are cost effective (there is no need for new infrastructure) and can take advantage of multiuser diversity (the more the users are the better relay nodes can be found). The main disadvantage of dynamic relays is the very high CSI feedback overhead needed for the relay selection and the selection complexity itself. The entity performing relay selection (BS) needs to acquire all the relevant CSI of the system in order to choose suitable relays for a specific transmission. In this dissertation we have devised ways of limiting this signaling demand by limiting the number of relay candidates for a specific transmission to a small but suitable subset of the cell users. This has been shown to greatly reduce feedback overhead and complexity without compromising performance.

Future Perspectives

The present dissertation has proposed some techniques for Multicell-MIMO and cooperative relaying. It is a step forward facilitating the practical implementation of these technologies and their inclusion to 4G standards, e.g. LTE-Advanced.

However there are still many aspects related to the aforementioned technologies that have not been addressed in this thesis. Throughout this dissertation the CSIT needed for scheduling and transmission has been assumed to arrive at the network infrastructure (at one or more BSs) through delayless links and usually without errors (only in Chapter 4 CSI is corrupted by errors introduced by the feedback links). These assumptions correspond to low mobility users whose CSI does not change rapidly; thus BSs have the opportunity to track these users' CSI. However in real wireless systems, a great percentage of users are of high mobility and thus the CSIT corresponding to them is usually outdated. As a result these scenarios need to be evaluated in order to get better insights on how mobility impacts Multicell-MIMO and Cooperative Relaying. Furthermore new techniques can be developed that allow users of higher mobilities to gain from MCP and relaying.

The assumption of perfect CSI, made in most of this thesis (apart from Chapter 4) is also unrealistic. In practical systems channel estimation is imperfect and also forms of limited CSI feedback, e.g. quantized feedback, need to be employed in order to keep signaling in affordable levels. Therefore it is important to devise methods for efficiently representing the channel state of an MCP scenario with a given number of bits. As we have only considered uncoded transmission, it is also necessary to investigate the FEC coding aspects related with MCP and cooperative relaying, especially in the multicarrier domain. Consequently the performance of MCP and relaying enabled systems needs to be optimized with respect to the aforementioned challenges.

Bibliography

[1] D. Martín-Sacristán, J. F. Monserrat, J. Cabrejas-Peñuelas, D. Calabuig, S. Garrigas, and N. Cardona, "On the way towards fourth-generation mobile: 3GPP LTE and LTE-Advanced," *EURASIP Journal on Wireless Communications and Networking*, vol. 2009, p. 10, Sept. 2009.

[2] M. Karakayali, G. Foschini, and R. Valenzuela, "Network coordination for spectrally efficient communications in cellular systems," *IEEE Wireless Communications Magazine*, vol. 13, no. 4, pp. 56–61, Aug. 2006.

[3] C. Botella, G. Pinero, A. Gonzalez, and M. De Diego, "Coordination in a multi-cell multi-antenna multi-user W-CDMA system: A beamforming approach," *IEEE Transactions on Wireless Communications*, vol. 7, no. 11, pp. 4479–4485, Nov. 2008.

[4] C. Botella, F. Domene, G. Pinero, and T. Svensson, "A low-complexity joint power control and beamforming algorithm for the downlink of multi-user W-CDMA coordinated systems," in *IEEE 10th Workshop on Signal Processing Advances in Wireless Communications, 2009. SPAWC '09.*, June 2009, pp. 226–230.

[5] A. Papadogiannis, E. Hardouin, and D. Gesbert, "Decentralising multicell cooperative processing: A novel robust framework," *EURASIP Journal on Wireless Communications and Networking*, vol. 2009, p. 10, Aug. 2009.

[6] S. Parkvall, E. Dahlman, A. Furuskar, Y. Jading, M. Olsson, S. Wanstedt, and K. Zangi, "LTE-Advanced - Evolving LTE towards IMT-Advanced," in *IEEE 68th Vehicular Technology Conference, 2008. VTC 2008-Fall.*, Calgary, Canada, Sept. 2008, pp. 1–5.

[7] Alcatel, "Collaborative MIMO for LTE-Advanced downlink," Alcatel Shanghai Bell and Alcatel Lucent, 3GPP TSG RAN WG1 Meeting 53bis, R1-082501, Warsaw, Poland, Tech. Rep., July 2008.

[8] Y. Song, L. Cai, K. Wu, and H. Yang, "Collaborative MIMO based on multiple base station coordination," contribution to IEEE 802.16m, IEEE C802.16m-07/162, Tech. Rep., Aug. 2007.

[9] H. Skjevling, D. Gesbert, and A. Hjørungnes, "Low-complexity distributed multibase transmission and scheduling," *EURASIP Journal on Advances in Signal Processing*, vol. 2008, p. 9 pages, 2008.

[10] S. Venkatesan, "Coordinating base stations for greater uplink spectral efficiency in a cellular network," in *IEEE 18th International Symposium on Personal, Indoor and Mobile Radio Communications, 2007. PIMRC '07*, Athens, Greece, Sept. 2007, pp. 1–5.

[11] ——, "Coordinating base stations for greater uplink spectral efficiency: Proportionally fair user rates," in *IEEE 18th International Symposium on Personal, Indoor and Mobile Radio Communications, 2007. PIMRC '07*, Athens, Greece, Sept. 2007, pp. 1–5.

[12] F. Boccardi and H. Huang, "Limited downlink network coordination in cellular networks," in *IEEE 18th International Symposium on Personal, Indoor and Mobile Radio Communications, 2007. PIMRC '07*, Athens, Greece, Sept. 2007, pp. 1–5.

[13] M. Kamoun and L. Mazet, "Base-station selection in cooperative single frequency cellular network," in *IEEE 8th Workshop on Signal Processing Advances in Wireless Communications, 2007. SPAWC '07.*, Helsinki, Finland, June 2007, pp. 1–5.

[14] A. Papadogiannis, D. Gesbert, and E. Hardouin, "A dynamic clustering approach in wireless networks with multi-cell cooperative processing," in *IEEE International Conference on Communications, 2008. ICC '08.*, Beijing, China, May 2008, pp. 4033–4037.

[15] F. Boccardi, H. Huang, and A. Alexiou, "Network MIMO with reduced backhaul requirements by MAC coordination," in *IEEE 42nd Asilomar Conference on Signals, Systems and Computers, 2008. ASILOMAR '08*, Pacific Grove, USA, Oct. 2008, pp. 1125–1129.

[16] C. B. Peel, B. M. Hochwald, and A. L. Swindlehurst, "A vector-perturbation technique for near-capacity multiantenna multiuser communication-part i: channel inversion and regularization," *IEEE Transactions on Communications*, vol. 53, no. 1, pp. 195–202, Jan. 2005.

[17] G. Dimic and N. Sidiropoulos, "On downlink beamforming with greedy user selection: performance analysis and a simple new algorithm," *IEEE Transactions on Signal Processing*, vol. 53, no. 10, pp. 3857–3868, Oct. 2005.

[18] F. Kaltenberger, M. Kountouris, D. Gesbert, and R. Knopp, "On the trade-off between feedback and capacity in measured MU-MIMO channels," *IEEE Transactions on Wireless Communications*, vol. 8, no. 9, pp. 4866–4875, Sept. 2009.

[19] F. Kaltenberger, D. Gesbert, R. Knopp, and M. Kountouris, "Performance of multi-user MIMO precoding with limited feedback over measured channels," in *IEEE Global Telecommunications Conference, 2008. GLOBECOM '08.*, New Orleans, USA, 30 Nov.-4 Dec. 2008, pp. 1–5.

[20] N. Jindal, "MIMO broadcast channels with finite-rate feedback," *IEEE Transactions on Information Theory*, vol. 52, no. 11, pp. 5045–5060, Nov. 2006.

[21] M. Kountouris, R. de Francisco, D. Gesbert, D. T. M. Slock, and T. Salzer, "Efficient metrics for scheduling in mimo broadcast channels with limited feedback," in *IEEE International Conference on Acoustics, Speech and Signal Processing, 2007. ICASSP '07.*, vol. 3, Apr. 2007, pp. III–109–III–112.

[22] T. S. Rappaport, *Wireless Communications - Principles and Practice*, 2nd ed. Prentice Hall, 2001.

[23] A. Osseiran, "Advanced antennas in wireless communications," Ph.D. dissertation, Royal Institute of Technology (KTH), Sweden, Apr. 2006.

[24] A. Goldsmith, *Wireless Communications.* Cambridge University Press, 2005.

[25] M.2133, "Requirements, evaluation criteria and submission templates for the development of IMT-Advanced," ITU-R, Tech. Rep., 2008.

[26] M.2134, "Requirements related to technical performance for IMT-Advanced radio interface(s)," ITU-R, Tech. Rep., 2008.

[27] M.2135, "Guidelines for evaluation of radio interface technologies for IMT-Advanced," ITU-R, Tech. Rep., 2008.

[28] TR36.913, "Requirements for further advancements for Evolved Universal Terrestrial Radio Access (E-UTRA) (LTE-Advanced)," 3GPP, Tech. Rep., 2008.

[29] TR36.912, "Feasibility study for further advancements for E-UTRA (LTE-Advanced)," 3GPP, Tech. Rep., 2009.

[30] TR36.814, "Evolved Universal Terrestrial Radio Access (E-UTRA); further advancements for E-UTRA physical layer aspects," 3GPP, Tech. Rep., 2009.

[31] S. Catreux, P. Driessen, and L. Greenstein, "Simulation results for an interference-limited Multiple-Input Multiple-Output cellular system," *IEEE Communications Letters*, vol. 4, no. 11, pp. 334–336, Nov. 2000.

[32] J. Andrews, "Interference cancellation for cellular systems: a contemporary overview," *IEEE Wireless Communications Magazine*, vol. 12, no. 2, pp. 19–29, Apr. 2005.

[33] R. Vaughan, "On optimum combining at the mobile," *IEEE Transactions on Vehicular Technology*, vol. 37, no. 4, pp. 181–188, Nov. 1988.

[34] D. Gesbert, S. Kiani, A. Gjendemsj, and G. ien, "Adaptation, coordination, and distributed resource allocation in interference-limited wireless networks," *Proceedings of the IEEE*, vol. 95, no. 12, pp. 2393–2409, Dec. 2007.

[35] S. Kiani, D. Gesbert, A. Gjendemsjo, and G. Oien, "Distributed power allocation for interfering wireless links based on channel information partitioning," *IEEE Transactions on Wireless Communications*, vol. 8, no. 6, pp. 3004–3015, June 2009.

[36] S. Kiani and D. Gesbert, "Optimal and distributed scheduling for multicell capacity maximization," *IEEE Transactions on Wireless Communications*, vol. 7, no. 1, pp. 288–297, Jan. 2008.

[37] S. Shamai and B. Zaidel, "Enhancing the cellular downlink capacity via co-processing at the transmitting end," in *IEEE 53rd Vehicular Technology Conference, 2001. VTC 2001-Spring.*, vol. 3, Rhodes, Greece, 2001, pp. 1745–1749 vol.3.

[38] H. Zhang and H. Dai, "Cochannel interference mitigation and cooperative processing in downlink multicell multiuser mimo networks," *EURASIP Journal on Wireless Communications and Networking*, vol. 2004, no. 2, p. 14, Feb. 2004.

[39] M. K. Karakayali, "Network coordination for spectrally efficient communications in wireless networks," Ph.D. dissertation, Rutgers, The State University of New Jersey, USA, Jan. 2007.

[40] G. Foschini, K. Karakayali, and R. Valenzuela, "Coordinating multiple antenna cellular networks to achieve enormous spectral efficiency," *IEE Proceedings on Communications*, vol. 153, no. 4, pp. 548–555, Aug. 2006.

[41] M. Karakayali, G. Foschini, R. Valenzuela, and R. Yates, "On the maximum common rate achievable in a coordinated network," in *IEEE International Conference on Communications, 2006. ICC '06*, vol. 9, Istanbul, Turkey, June 2006, pp. 4333–4338.

[42] O. Somekh, O. Simeone, Y. Bar-Ness, and A. Haimovich, "Distributed multi-cell zero-forcing beamforming in cellular downlink channels," in *IEEE Global Telecommunications Conference, 2006. GLOBECOM '06*, San Francisco, USA, 27 Nov.-1 Dec. 2006, pp. 1–6.

[43] S. Jing, D. N. C. Tse, J. B. Soriaga, J. Hou, J. E. Smee, and R. Padovani, "Downlink macro-diversity in cellular networks," in *IEEE International Symposium on Information Theory, 2007. ISIT '07*, Nice, France, June 2007, pp. 1–5.

[44] S. Jing, D. N. C. Tse, J. Hou, J. B. Soriaga, J. E. Smee, and R. Padovani, "Multi-cell downlink capacity with coordinated processing," in *Information Theory and Applications Workshop, 2007. ITA '07*, San Diego, USA, Jan. 2007.

[45] P. Marsch and G. Fettweis, "A framework for optimizing the downlink of distributed antenna systems under a constrained backhaul," in *European Wireless Conference, 2007. EW '07*, Paris, France, Apr. 2007.

[46] ——, "A framework for optimizing the uplink performance of distributed antenna systems under a constrained backhaul," in *IEEE International Conference on Communications, 2007. ICC '07*, Glasgow, UK, June 2007, pp. 975–979.

[47] ——, "On multicell cooperative transmission in backhaul-constrained cellular systems," *Springer Annals of Telecommunications*, vol. 63, no. 5-6, pp. 253–269, May 2008.

[48] A. Sendonaris, E. Erkip, and B. Aazhang, "User cooperation diversity. part i. system description," *IEEE Transactions on Communications*, vol. 51, no. 11, pp. 1927–1938, Nov. 2003.

[49] J. N. Laneman and G. W. Wornell, "Distributed space-time-coded protocols for exploiting cooperative diversity in wireless networks," *IEEE Transactions on Information Theory*, vol. 49, no. 10, pp. 2415–2425, Oct. 2003.

[50] J. N. Laneman, D. N. C. Tse, and G. W. Wornell, "Cooperative diversity in wireless networks: Efficient protocols and outage behavior," *IEEE Transactions on Information Theory*, vol. 50, no. 12, pp. 3062–3080, Dec. 2004.

[51] R. Nabar, H. Bolcskei, and F. Kneubuhler, "Fading relay channels: performance limits and space-time signal design," *IEEE Journal on Selected Areas in Communications*, vol. 22, no. 6, pp. 1099–1109, Aug. 2004.

[52] R. Pabst, B. Walke, D. Schultz, P. Herhold, H. Yanikomeroglu, S. Mukherjee, H. Viswanathan, M. Lott, W. Zirwas, M. Dohler, H. Aghvami, D. D. Falconer, and G. P. Fettweis, "Relay-based deployment concepts for wireless and mobile broadband radio," *IEEE Communications Magazine*, vol. 42, no. 9, pp. 80–89, Sept. 2004.

[53] A. Bletsas, A. Khisti, D. Reed, and A. Lippman, "A simple cooperative diversity method based on network path selection," *IEEE Journal on Selected Areas in Communications*, vol. 24, no. 3, pp. 659–672, Mar. 2006.

[54] A. Bletsas, H. Shin, and M. Win, "Cooperative communications with outage-optimal opportunistic relaying," *IEEE Transactions on Wireless Communications*, vol. 6, no. 9, pp. 3450–3460, Sept. 2007.

[55] A. Alexiou and F. Boccardi, "Coordination and cooperation for next generation wireless systems- overhead signalling requirements and cross layer considerations," *IEEE International Conference on Acoustics, Speech, and Signal Processing, 2009. ICCASP '09*, pp. 3613–3616, 2009.

[56] D. Gesbert, M. Shafi, D. shan Shiu, P. Smith, and A. Naguib, "From theory to practice: an overview of MIMO space-time coded wireless systems," *IEEE Journal on Selected Areas in Communications*, vol. 21, no. 3, pp. 281–302, Apr. 2003.

[57] E. Biglieri, R. Calderbank, A. Constantinides, A. Goldsmith, A. Paulraj, and H. V. Poor, *MIMO Wireless Communications.* Cambridge University Press, 2007.

[58] M. Pischella and J.-C. Belfiore, "Distributed resource allocation for rate-constrained users in multi-cell OFDMA networks," *IEEE Communications Letters*, vol. 12, no. 4, pp. 250–252, Apr. 2008.

[59] ——, "Power control in distributed cooperative OFDMA cellular networks," *IEEE Transactions on Wireless Communications*, vol. 7, no. 5, pp. 1900–1906, May 2008.

[60] ——, "Resource allocation for QoS-aware OFDMA using distributed network coordination," *IEEE Transactions on Vehicular Technology*, vol. 58, no. 4, pp. 1766–1775, May 2009.

[61] G. Caire and S. Shamai, "On the achievable throughput of a multi-antenna gaussian broadcast channel," *IEEE Transactions on Information Theory*, vol. 49, no. 7, pp. 1691–1706, July 2003.

[62] H. Weingarten, Y. Steinberg, and S. Shamai, "The capacity region of the gaussian Multiple-Input Multiple-Output broadcast channel," *IEEE Transactions on Information Theory*, vol. 52, no. 9, pp. 3936–3964, Sept. 2006.

[63] Q. Spencer, C. Peel, A. Swindlehurst, and M. Haardt, "An introduction to the multi-user mimo downlink," *IEEE Communications Magazine*, vol. 42, no. 10, pp. 60–67, Oct. 2004.

[64] M. Kountouris, "Multiuser multi-antenna systems with limited feedback," Ph.D. dissertation, Télécom Paris (ENST), France, Jan. 2008.

[65] R. de Francisco, "Performance optimization of MIMO systems with partial channel state information," Ph.D. dissertation, Télécom Paris (ENST), France, Dec. 2007.

[66] D. Gesbert and M.-S. Alouini, "How much feedback is multi-user diversity really worth?" in *IEEE International Conference on Communications, 2004. ICC '04*, vol. 1, Paris, France, June 2004, pp. 234–238.

[67] V. Hassel, D. Gesbert, M.-S. Alouini, and G. Oien, "A threshold-based channel state feedback algorithm for modern cellular systems," *IEEE Transactions on Wireless Communications*, vol. 6, no. 7, pp. 2422–2426, July 2007.

[68] X. Qin and R. Berry, "Opportunistic splitting algorithms for wireless networks," in *23rd Annual Joint Conference of the IEEE Computer and Communications Societies, 2004. INFOCOM '04.*, vol. 3, Mar. 2004, pp. 1662–1672 vol.3.

[69] ——, "Opportunistic splitting algorithms for wireless networks with fairness constraints," in *IEEE 4th International Symposium on Modeling and Optimization in Mobile, Ad Hoc and Wireless Networks, 2006*, Boston, USA, Apr. 2006, pp. 1–8.

[70] A. Papadogiannis, H. Bang, D. Gesbert, and E. Hardouin, "Downlink overhead reduction for multi-cell cooperative processing enabled wireless networks," in *IEEE 19th International Symposium on Personal, Indoor and Mobile Radio Communications, 2008. PIMRC 2008.*, Cannes, France, Sept. 2008, pp. 1–5.

[71] Y. W. Hong, W. J. Huang, F. H. Chiu, and C. C. J. Kuo, "Cooperative communications in resource-constrained wireless networks," *IEEE Signal Processing Magazine*, vol. 24, no. 3, pp. 47–57, May 2007.

[72] E. Yilmaz, R. Knopp, and D. Gesbert, "Some systems aspects regarding compressive relaying with wireless infrastructure links," in *IEEE Global Telecommunications Conference, 2008. IEEE GLOBECOM 2008.*, New Orleans, USA, 30 Nov.-4 Dec. 2008, pp. 1–5.

[73] ——, "On the gains of fixed relays in cellular networks with intercell interference," in *IEEE 10th Workshop on Signal Processing Advances in Wireless Communications, 2009. SPAWC '09.*, Perugia, Italy, June 2009, pp. 603–607.

[74] E. Yilmaz, F. Boccardi, and A. Alexiou, "Distributed and centralized architectures for relay-aided cellular systems," in *IEEE 69th Vehicular Technology Conference, 2009. VTC Spring-2009.*, Barcelona, Spain, Apr. 2009, pp. 1–5.

[75] A. Wolfgang, N. Seifi, and T. Ottosson, "Resource allocation and linear precoding for relay assisted multiuser MIMO systems," in *International ITG Workshop on Smart Antennas, 2008. WSA '08.*, Darmstadt, Germany, Feb. 2008, pp. 162–168.

[76] S. Song, K. Son, H.-W. Lee, and S. Chong, "Opportunistic relaying in cellular network for capacity and fairness improvement," in *IEEE Global Telecommunications Conference, 2007. GLOBECOM '07.*, Nov. 2007, pp. 4407–4412.

[77] G. Calcev and J. Bonta, "Opportunistic two-hop relays for ofdma cellular networks," in *IEEE Global Telecommunications Conference Workshops, 2008. GLOBECOM '08*, 30 Nov.-4 Dec. 2008, pp. 1–6.

[78] A. Papadogiannis, E. Hardouin, A. Saadani, D. Gesbert, and P. Layec, "A novel framework for the utilisation of dynamic relays in cellular networks," in *IEEE 42nd Asilomar Conference on Signals, Systems and Computers, 2008. ASILOMAR '08*, Pacific Grove, USA, Oct. 2008, pp. 975–979.

[79] A. Papadogiannis, A. Saadani, and E. Hardouin, "Exploiting dynamic relays with limited overhead in cellular systems," in *IEEE Global Telecommunications Conference Workshops, 2009. GLOBECOM '09*, Hawaii, USA, 30 Nov.-4 Dec. 2009.

[80] H. Viswanathan and S. Mukherjee, "Performance of cellular networks with relays and centralized scheduling," *IEEE Transactions on Wireless Communications*, vol. 4, no. 5, pp. 2318–2328, Sept. 2005.

[81] TR25.814, "Physical layer aspect for the Evolved Universal Terrestrial Radio Access (UTRA)," 3GPP, Tech. Rep., 2006.

[82] U. Charash, "Reception through nakagami fading multipath channels with random delays," *IEEE Transactions on Communications*, vol. 27, no. 4, pp. 657–670, Apr. 1979.

[83] M. Nakagami, "The m-distribution - A general formula of intensity distribution of rapid fading," in *Statistical Methods in Radio Wave*

Propagation, W. G. Hoffman, Ed. Oxford, UK: Permagon Press, 1960, pp. 3–36.

[84] I. S. Gradshteyn and I. M. Ryzhik, *Table of Integrals, Series, and Products*, 6th ed. Academic, 2000.

[85] J. G. Proakis, *Digital Communications*, 3rd ed. McGraw-Hill, 1995.

[86] M. Lötter and P. van Rooyen, "Cellular channel modeling and the performance of ds-cdma systems with antenna arrays," *IEEE Journal on Selected Areas in Communications*, vol. 17, no. 12, pp. 2181–2196, Dec. 1999.

[87] M. Abdel-Hafez and M. Safak, "Performance analysis of digital cellular radio systems in Nakagami fading and correlated shadowing environment," *IEEE Transactions on Vehicular Technology*, vol. 48, no. 5, pp. 1381–1391, Sept. 1999.

[88] D. Tse and P. Viswanath, *Fundamentals of Wireless Communication*. Cambridge University Press, 2005.

[89] T. Sälzer, "Transmission strategies employing multiple antennas for the downlink of multi-carrier CDMA systems," Ph.D. dissertation, Institut National des Sciences Appliqués de Rennes, France, June 2004.

[90] T. Yoo and A. Goldsmith, "On the optimality of multiantenna broadcast scheduling using zero-forcing beamforming," *IEEE Journal on Selected Areas in Communications*, vol. 24, no. 3, pp. 528–541, Mar. 2006.

[91] S. Jafar and S. Srinivasa, "On the optimality of beamforming with quantized feedback," *IEEE Transactions on Communications*, vol. 55, no. 12, pp. 2288–2302, Dec. 2007.

[92] F. Boccardi and H. Huang, "Zero-forcing precoding for the MIMO broadcast channel under per-antenna power constraints," in *IEEE 7th Workshop on Signal Processing Advances in Wireless Communications, 2006. SPAWC '06.*, July 2006, pp. 1–5.

[93] A. Papadogiannis, E. Hardouin, and D. Gesbert, "A framework for decentralising multi-cell cooperative processing on the downlink," in *IEEE Global Telecommunications Conference Workshops, 2008. GLOBECOM '08*, New Orleans, USA, 30 Nov.-4 Dec. 2008, pp. 1–5.

[94] CELTIC, "D1.4 initial report on advanced multiple antenna systems," WINNER+ deliverable, Available online: http://projects.celtic-initiative.org/winner+/deliverables_winnerplus.html, Tech. Rep., Jan. 2009.

[95] Alcatel, "LS on support of ACK/NACK repetition in Rel-8," 3GPP TSG RAN WG1 meeting, Tech. Rep., November 2008.

[96] D. Gesbert, M. Kountouris, R. Heath, C.-B. Chae, and T. Salzer, "Shifting the MIMO paradigm," *Signal Processing Magazine, IEEE*, vol. 24, no. 5, pp. 36–46, Sept. 2007.

[97] D. J. Love, R. W. Heath, V. K. N. Lau, D. Gesbert, B. D. Rao, and M. Andrews, "An overview of limited feedback in wireless communication systems," *IEEE Journal on Selected Areas in Communications*, vol. 26, no. 8, pp. 1341–1365, Oct. 2008.

[98] W. Choi and J. Andrews, "The capacity gain from intercell scheduling in multi-antenna systems," *IEEE Transactions on Wireless Communications*, vol. 7, no. 2, pp. 714–725, Feb. 2008.

[99] Y. B. Gol'dshtein, "Portrait of the inverse of a sparse matrix," *Journal on Cybernetics and Systems Analysis*, vol. 28, no. 4, pp. 514–519, July 1992.

[100] A. Papadogiannis and G. Alexandropoulos, "System level performance evaluation of dynamic relays in cellular networks over nakagami-m fading channels," in *IEEE 20th International Symposium on Personal, Indoor and Mobile Radio Communications, 2009. PIMRC '09.*, Tokyo, Japan, Sept. 2009.

[101] K. Woradit, T. Quek, W. Suwansantisuk, H. Wymeersch, L. Wuttisittikulkij, and M. Z. Win, "Outage behavior of cooperative diversity with relay selection," in *IEEE Global Telecommunications Conference, 2008. IEEE GLOBECOM '08.*, New Orleans, USA, 30 Nov.-4 Dec. 2008, pp. 1–5.

[102] I. F. Akyildiz and I. H. Kasimoglu, "Wireless sensor and actor networks: research challenges," *ELSEVIER Ad Hoc Networks*, vol. 2, no. 4, pp. 351 – 367, 2004.

[103] M. Munir, A. Papadogiannis, and F. Filali, "Cooperative multi-hop wireless sensor-actuator networks: Exploiting actuator cooperation

and cross-layer optimizations," in *IEEE Wireless Communications and Networking Conference, 2008. WCNC '08.*, Las Vegas, USA, 31 Mar.-3 Apr. 2008, pp. 2881–2886.

www.ingramcontent.com/pod-product-compliance
Lightning Source LLC
Chambersburg PA
CBHW021053210326
41598CB00016B/1203